CW00504775

Riemann Hypothesis and spectral theory

Jason Cole

This publication is in copyright. ©

Table of contents

1. <u>Introduction</u>

This book provides a brief overview of the Riemann Zeta function, Riemann Hypothesis and the Hilbert-Polya spectral operator approach to proving RH. Also included in this book is a new discovery that describes a correlation between the Riemann Xi function and gravity rotational curves. Surprisingly their is a mathematical correlation between the complex system of the Riemann Xi function and the large scale distribution of galaxies and rotational curves. Also included in this book are new discoveries on the Prime Number theorem(PNT), Riemann Zeta function and other new science and math discoveries.

2. New research on Zeta that will lead towards a RH proof

Please note that the information in this book is research concerning the study of a new form of the Riemann Zeta function. It is "NOT" written as a proof of Riemann Hypothesis but offers mathematicians new insight to help formulate a rigorous proof of the Riemann Hypothesis. This book provides mathematicians with new tools and insights to help formulate a proof for RH(give it a good start). Over the past 160 yrs many mathematicians and some Physicists have present proofs to the Riemann Hypothesis but they all have failed. The reason why in my opinion isn't a lack of ingenuity of the Mathematicians (or Physicists) to derive a proof. It was simply not enough understanding about the behavior of the Riemann Zeta function for them to come up with a proof. If they didn't fully understand why the nontrivial zeros form on the critical line and what factors in the functions causes the critical line to form then its difficult or virtually impossible to prove all nontrivial zeros have a real part equal to 1/2.

Proving the Riemann Hypothesis using pure mathematics is one approach. The other approach to proving the Riemann Hypothesis is by Physics. Their is mysterious connection between the spacing of the nontrivial zeros of Zeta to the energy level spacing inside heavy atomic nuclei.

What this book provides is new and fresh ideas on how to attach the Riemann Hypothesis that hasn't been explored before. Based on the research in this book it could lead towards a proof or one of many different types of proof for the Riemann Hypothesis.

The following chapter provides a brief history of the Riemann Zeta function. Riemann Hypothesis is based on proving a certain aspect of the Riemann Zeta function.

3. <u>A brief description of the Riemann Zeta function</u>

In 1720's Euler created the Zeta function

$$\zeta(s) = 1 + \frac{1}{2^s} + \frac{1}{3^s} + \frac{1}{4^s} + \frac{1}{5^s} + \frac{1}{6^s} + \frac{1}{7^s} + \cdots$$

What made this infinite series so interesting is that it has a connection to Prime numbers. Meaning if you factor this infinite series you have a infinite product over the Primes,

$$\sum_n \frac{1}{n^s} = \prod_p \frac{1}{1 - p^{-s}}$$

In the 1800s Bernhard Riemann came along and extended the Zeta function into the complex plane. Rather than just using real variables like Euler, he use complex numbers in the form of s=x+yi. Euler only considered the Zeta function for values greater than one but Riemann extended the Zeta function for all values including those for less than 1.

Using very complicated mathematical opera-
tors, Riemann transformed the original Euler Zeta function into a complex number Zeta
function. From that point on the Zeta function was called the Riemann Zeta function.
Below is a equation of his famous functional
equation. It not only includes the same values as the original Zeta function greater than

$$\zeta(s) = 2^s \pi^{s-1} \sin\left(\frac{\pi s}{2}\right) \Gamma(1-s)\zeta(1-s)$$

than one it includes all values less than one
and all complex numbers. When Euler created the Zeta function it equaled a product over
the Primes. When Riemann extended this
Zeta function to complex number he discovered a mysterious line in the function called
the "critical line". On this critical line as

shown in the graph above its roots(zeros) has
a new connection to the distribution of Prime
Numbers. The following is a graph of these
nontrivial zeros on the critical line.

5

Below is a picture of the Zeta function and its nontrivial zeros on the vertical line and the trivial zeros on the horizontal line.

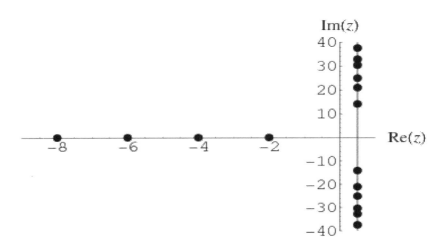

The trivial zeros (-2,-4,-6...) on the horizontal line are called trivial because they appear ordinary and has no connection to the distribution of Prime numbers. However, on the critical line (imaginary axis) the distribution of those zeros called nontrivial zeros are of interest.

Riemann created a equation that count the number of Primes less than a given number. What made the equation so unique and powerful is that it used the nontrivial zeros on the critical line of the Zeta function to subtract and correct the Prime estimates from the Prime counting function Li(x).

$$J(x) = Li(x) - \sum_{p} Li(x^p) - \log 2 + \int_{x}^{\infty} \frac{1}{t(t^2 - 1)\log t} dt$$

In order for this Prime counting function to be true all the nontrivial zeros had to be in the form of 1/2+ yi. All the nontrivial zeros (roots) had to have a real part equal to 1/2. This conjecture that all the nontrivial zeros having real part equal to one half is called the Riemann hypothesis.

4. <u>Race for a proof of Riemann hypothesis</u>

Ever since Bernhard Riemann conjectured the proof of the Riemann hypothesis, numerous mathematicians have tried to proposed a proof. Also the nontrivial zeros on the critical line have be computed up to 10^{12} zero. Meaning up to that height all the nontrivial zeros still lie on the line. Some mathematicians suggest that a nontrivial zero may be off the line up to a huge number. However, that can number be reached by a computer. Mathematicians have tried to proposed proofs that all the nontrivial zeros only lie on the critical line. In which they tried to show that no other nontrivial zero is off the critical and no other nontrivial zeros exist off the critical line throughout all infinity. Their proofs as of the writing of this book has failed.

Beside looking for a proof, Mathematicians have made tremendous progress in Number Theory showing new mathematical discoveries based on "if Riemann Hypothesis is true". Countless mathematical theorems are based on whether Riemann Hypothesis is true or not.

When you think about it, trying to find a proof of all the nontrivial zeros being only on the critical line is no easy task. You can prove their an infinite number of nontrivial zeros on the critical line (someone has done that) but their could be one or two or an infinite number of nontrivial zeros off the critical line.

What made finding a proof of RH (Riemann Hypothesis) even more exciting is that the nontrivial zeros on the critical line has a interesting connection to physics. For instance, based on a conjecture called the "Montgomery Pair correlation conjecture" the spacing (distance) between the nontrivial zeros on the critical line is statistically equivalent to the spacing of eigenvalues in a GUE operator in Random Matrix Theory. The following page shows the graph of how the spacing of the nontrivial zeros on the critical line of the Riemann Zeta function correspond to the spacing of eigenvalues in a Random Matrix. That correspondence supports the Hilbert-Polya conjecture to prove RH.

5. <u>The Hilbert-Polya conjecture (Spectral theory)</u>

The Hilbert-Polya operator is defined as the nontrivial zeros of the Riemann Zeta function corresponding to the eigenvalues of a some Hermitian operator.

Evidence to support the Hilbert-Polya conjecture comes in the form of the the Montgomery Pair conjecture(graph below).

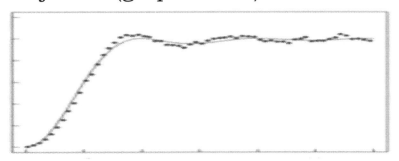

From the Montgomery pair correlation graph above the spacing between pairs of nontrivial zeros of the Riemann Zeta function correspond on a statistical level to the pair correlation of energy levels in a GUE operator. Meaning the spacing of energy levels inside heavy atomic nuclei correspond on a statistical level to the spacing between the nontrivial zeros on the critical line of the Zeta function.

Finding a physics operator to prove Riemann Hypothesis (all nontrivial zeros of the Riemann Zeta function have a real part equal to 1/2) would be more exciting than a pure Number Theory approach to proving RH.

Complex Random matrix theory & RH
(The full picture of the Montgomery pair correlation conjecture)

The proof for the Riemann Hypothesis may come by a Number Theory approach or a physics operator approach such as the Hilbert-Polya spectral operator approach to proving RH. The following chapter provides new insight into the Montgomery pair correlation conjecture that provides a glimpse of the type of Physics operator RH may be based upon.

6. New insight into the Montgomery pair correlation conjecture

The Montgomery pair correlation conjecture describes a correlation between the spacing of the nontrivial zeros of the Riemann Zeta function to the spacing of energy level inside heavy atomic nuclei. The graph highlights the overlapping correlation behavior between Zeta zeros and quantum energy levels.

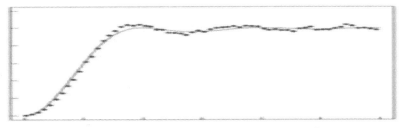

The nontrivial zeros of the Zeta function has the same nontrivial zeros as the Riemann Xi function. The difference is that the Riemann Xi function nontrivial zeros are purely real.

R. Zeta function   Riemann Xi function

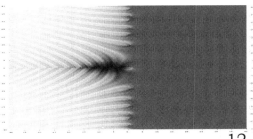 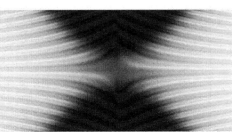

12

New GUE discovery(Complex Random matrix)

If we input complex variables into a GUE operator then it's graph looks identical to a complex normal distribution curve & Xi.

$$\xi(s) = \xi(1 - s)$$

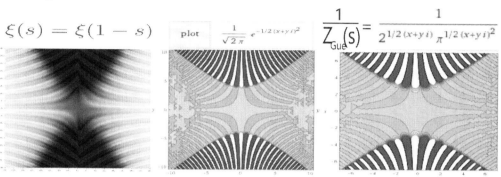

$$\text{plot} \quad \frac{1}{\sqrt{2\pi}} e^{-1/2\,(x+y\,i)^2} \qquad \frac{1}{Z_{Gue}(s)} = \frac{1}{2^{1/2\,(x+y\,i)}\,\pi^{1/2\,(x+y\,i)^2}}$$

Complex GUE operator equation above(graph on right) looks like the Riemann Xi function. This GUE has real zeros like the Xi function.

$$\xi(s) \sim \frac{1}{Z_{GUE(s)}} e^{-\frac{s}{2}\mathrm{tr}H^2}$$

The following graph highlights the Montgomery pair correlation conjecture in the C. plane

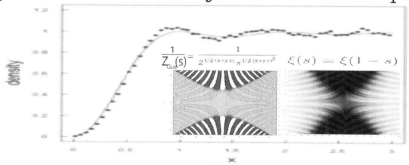

13

7. **Rotational curve correlation conjecture**

Montgomery Pair correlation conjecture describes the repulsive spacing between the nontrivial zeros matching the repulsive spacing between the energy levels inside heavy atomic nuclei. This chapter shows a new correlation with a galaxy rotational curve.

<u>Pair correlation</u>   <u>Flat rotation curve</u>

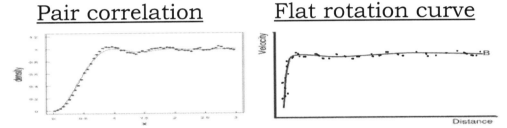

The flat rotational curve is based on a repulsion effect similar to energy level repulsion. The 2 graphs above have repulsion. Dark matter has anti-gravity with a r^2 curve that overlaps the 1/r^2 curve. The overlap of r^2/r^2=1 causes the flat rotational curve.

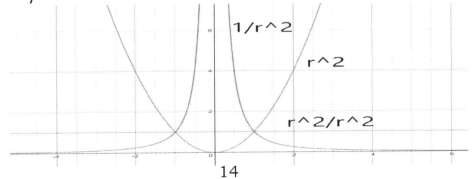

14

The 3 rotational curves can also be applied to the universe. The overlap of a deflationary $1/x^2$ curve & inflationary x^2 curve causes the $x^2/x^2=1$ flat geometry universe.

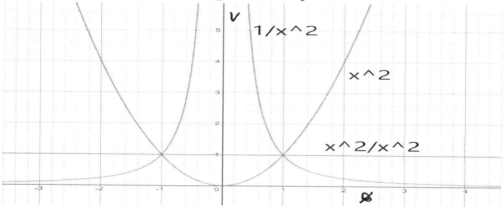

The r^2 rotational curve or x^2 Chaotic inflation curve are ghost curves based on Tachyon (negative mass squared) energy(mass). The origin of the imaginary energy(mass) come from a complex energy big bang. We had a complex conjugate pair big bang $(a+bi)(a-bi)$ rather than a matter-antimatter big bang. A complex energy universe implies a QM superposition complex wave-function (multiverse). The irony with Special Relativity is that the unwillingness to accept Tachyons (imag. variables) like QM uses complex numbers may be the reason GR & QM can't be combined.

15

The new Keplerian rotation curves

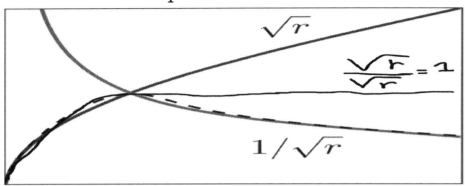

<u>What causes the (√r) curve for galaxies?</u>
The √r rotation curve can't be normal gravity curve because it would add to the 1/√r
curve and pull everything towards the center.
The √r rotation curve is based on anti-gravity and it is balancing out the normal gravity
curve for a flat curve. Because a galaxy has
a black hole in it unlike a Solar system, the
mysterious interior of a black hole may be the
driving force behind the √r rotational curve.
Below the event horizon, space falls FTL. Possibly the FTL geodesic physics in the interior of black hole (like Tachyons for Special
R.) causes the external √r rotation curve that
balances out 1/√r for a flat curve √r/√r = 1.
FTL space-flow reverse inside a blackhole like
a Tachyon appear in FTL Special Relativity.

The Rotational curve correlation conjecture

The take away from the new Keplerian curve
is that the flat rotational curve of $\sqrt{r}/\sqrt{r} = 1$
uses repulsive gravity. It's graph behaves like
the Montgomery Pair correlation conjecture
that also uses repulsion but between energy
levels or Zeta zeros. The systems are similar.

<u>Flat rotation curve</u>   <u>Pair correlation</u>

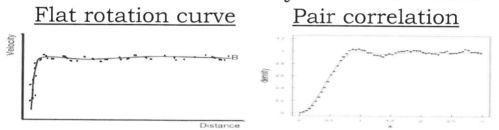

On a QM level a flat rotational curve has re-
pelling energy levels like a GUE random matrix. A solar system energy levels don't repel.

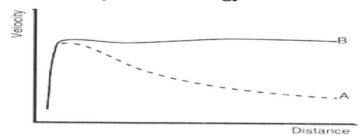

From a quantum mechanical perspective the
graphs of flat rotational curves behave like
the graphs in Random Matrix theory(RMT).
This may be a glimpse into Quantum gravity.

The Rotational curve correlation conjecture

The flat rotational curve conjecture has connection to the Montgomery pair correlation were it hints at a Quantum gravity system. If we compare the Pair correlation conjecture graph to the flat rotational curve graph they look similar. Also both work off repulsion. The nontrivial zeros(or energy levels) have repulsion & the Tachyon r^2 ghost curve(anti-gravity) overlaps the normal curve $1/r^2$ and create a flat rotational curve $r^2/r^2=1$.

Pair correlation	Flat rotation curve

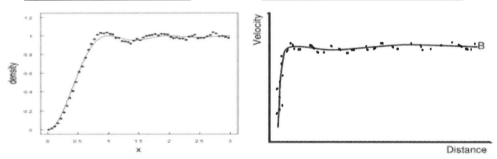

The physics of each system are not identical but they have parallels in how they behave on different size scales. The Tachyon r^2 rotational ghost curve is causing repulsion in the $1/r^2$ decreasing curve for a flat rotational curve analogous to how the atomic energy level repel each other for a flat curve graph.

18

A similar correlation shows up in a different
Zeta function with a rotational system. For
example, lets compare the Riemann Xi function to rotational curves and their complex
pair mirror image curves:(1/r^2,r^2,r^2/
r^2, -1/r^2, -r^2, -r^2/r^2). The Riemann Xi
graph looks similar to the 6 rotational curves.

<u>Six Rotational curves</u>   <u>Riemann Xi function</u>

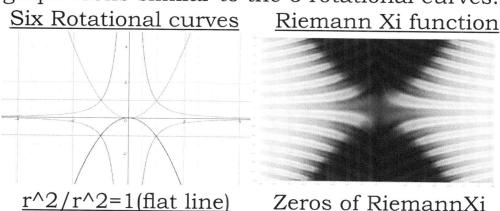

<u>r^2/r^2=1(flat line)</u>   <u>Zeros of RiemannXi</u>
The -/+ R. Xi zeros
are on the horizontal
axis like the flat
rotational curve above.

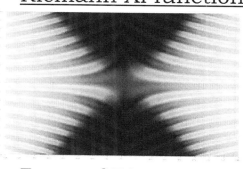

The <u>Rotational curve correlation conjecture</u>
parallels the Montgomery pair correlation.

<u>Flat rotation curve</u>   <u>Pair correlation</u>

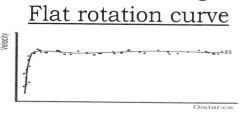   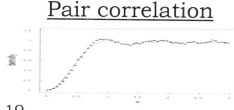

19

Complex normal distribution curve and gravity rotational curves

The 6 different rotation curves are related to a complex normal distribution curve graph.

| Distribution curve | Six Rotational curves |

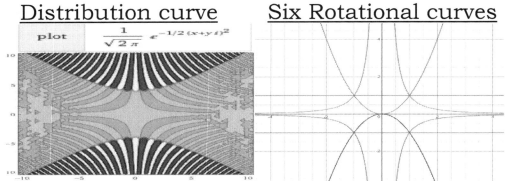

The six rotational curves and their complex conjugate pair mirror image curves are $1/r^2, r^2, r^2/r^2, -1/r^2, -r^2, -r^2/r^2$
Also the six rotational curves correlate to a six type inflationary universe model(x^2, x^2/x^2, $1/x^2, -x^2, -x^2/x^2, -1/x^2$). Their complex conjugate pair inflation curves. The overall 6 different universe curve models follows a complex normal distribution curve.

Six universe curves

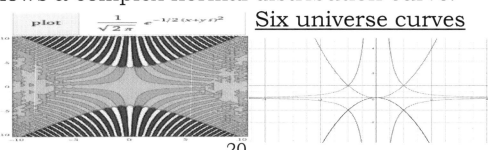

<u>A Complex number system universe</u>
The overall distribution of the universe and
gravity rotational curves follows a normal distribution in the complex plane! It even has
correlation to the Complex graph Riemann Xi
function(it's zeros related to energy levels).
Riemann Xi   plot $\frac{1}{\sqrt{2\pi}}e^{-1/2(x+yi)^2}$ Gravity curves

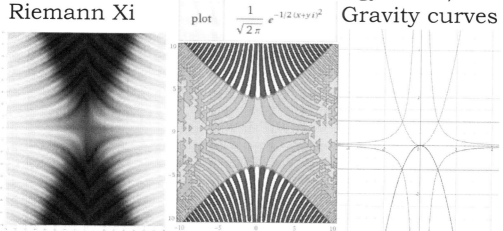

The Riemann Xi is a complex graph. For the
Gravity rotational curve and Gravity inflationary curves the the positive and negative parabola correlations to tachyon FTL imaginary
energy(matter). The decreasing curves corresponds to real energy(matter). Together they
make up a Complex system like Riemann Xi.
The Rotational curve correlation conjecture
points towards a Complex conjugate pair big
bang(QM complex wavefunctions: multiverse).

Real eigenvalues vs complex eigenvalues

When we compare the graph of the Riemann Xi function to the 6 rotational curves
its important to note the eigenvalues of the
Riemann Xi function are real. It's only one
horizontal line of eigenvalues on the x-axis.
However, when we look at the complex rotational curve graph we see complex conjugate
critical horizontal flat curves.

Riemann Xi	6 Gravity rotation curves
(Real eigenvalues)	(Complex eigenvalues)

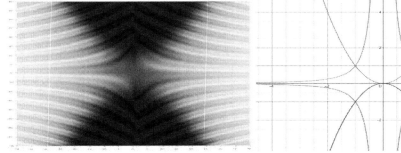

The Riemann Xi function has real eigenvalues. Their is only one horizontal line of real
eigenvalues on the x-axis. However, we have
complex conjugate pair flat rotational curves
on horizontal lines (r^2/r^2=1 and -r^2/r^2=
-1). That represent eigenvalues as complex.
Imagine a complex normal distribution

curve with complex eigenvalues.
The key difference in the "Rotational curve correlation conjecture" is that the two models behave similarity except the eigenvalues for the Riemann Xi function are real but the eigenvalues for the complex rotational curve are complex. That is why we have complex conjugate pair flat rotational curves. Those complex conjugates curves could be multiplied together to give a real value:
$(a+bi)(a-bi) = a^2+b^2$.

Random Matrix gravity(RMG)

Quantum gravity as discrete eigenvalues

When we look at the complex normal distribution curve of gravity rotational systems the discreteness appears to be in the flat rotational curve similar to the Montgomery pair correlation curve. Meaning for the flat rotational curve of galaxies it's possible it could be made of discrete eigenvalues(energy levels) of gravity like Zeta nontrivial zeros or eigenvalues levels in RMT. Based on $E=mc^2$ those discrete gravity eigenvalues(energy levels) could <u>equal particles that could be detected in a particle accelerator</u>.

Both flat graph curves stops at "1"

If you look at the height of the flat rotation-al curve model it stops at "1" were r^2/r^2=1 and if you look at the height of the Montgom-ery pair correlation conjecture it also stops at "1" based on the following equation:

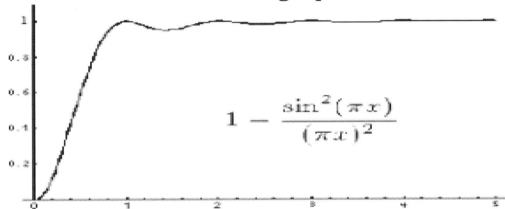

$$1 - \frac{\sin^2(\pi x)}{(\pi x)^2}$$

Let's compare the graphs side by side and notice how each stops at "1" for their height. Both are related to a normalizing constant.

$\sqrt{R}/\sqrt{R} = 1$ Height 1 for density

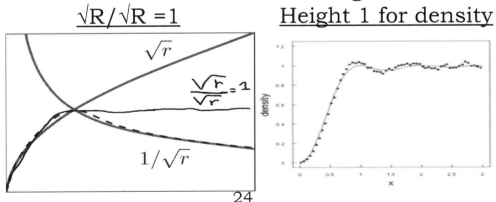

24

We see a <u>correlation</u> between a RMT (Random Matrix Theory) of energy levels bounded inside a heavy atomic nuclei and energy levels in a bounded gravitational galaxy system(bounded by the central black hole).

<u>Bounded gravity-energy levels</u>

If the following graphs of the

<u>Flat rotation curve</u>   <u>Pair correlation</u>

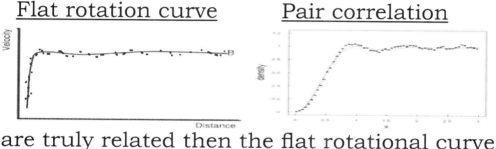

are truly related then the flat rotational curve for galaxies can be described by <u>discrete energy levels</u> in level repulsion just like the level repulsion in energy levels or Zeta zeros in the Montgomery pair correlation. The visible matter in the flat rotation curves for galaxies are following quantum energy level repulsion. Just as quantum systems are bound and force quantized energy states, gravity in a rotational system acts as a bound and forces rotational energy to be quantized. We see that in the flat rotational curve that looks like the Mont. pair correlation conjecture curve.

Rotational quantized energy of galaxies

According to Wikipedia
"A quantum mechanical system or particle
that is bound—that is, confined spatially—
can only take on certain discrete values of energy, called energy levels. We can also have
rotational energy levels in molecules. The energy spectrum of a system with such discrete
energy levels is said to be quantized"

What the "Rotational curve correlation conjecture" proposes is that a quantum mechanical system can be bound---that is, confined
spatially-- by gravity can only take on discrete values of energy levels. That flat rotational curve graph is due to quantized energy
levels like rotational energy levels in molecules.

Flat rotation curve   Pair correlation

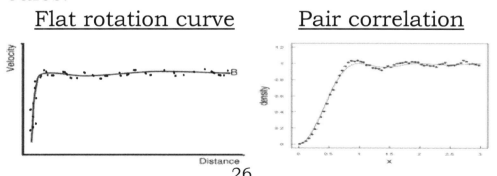

Our Solar system rotation system is also
bound by gravity like a galaxy rotation curve.
The difference is that the flat rotational curve
energy levels repel while the rotational energy
levels in Solar systems do not. Both types of
curves can be represented by Random matrix
curves.

Rotational curves   Pair correlation

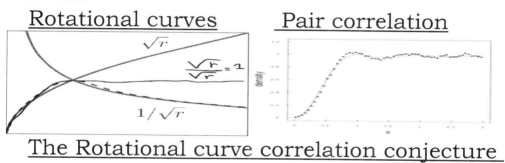

The Rotational curve correlation conjecture
is a glimpse into Quantum Gravity

This gravity "Rotational curve correlation
conjecture" model with its 3 different types of
rotational curves: $1/\sqrt{r}$, $\sqrt{r}/\sqrt{r}$, $\sqrt{r}$, has amazing parallels to Random Matrix Theory in
the Montgomery pair correlation conjecture.
Both the flat rotational curve and Pair correlation both work on repulsion. The gravity
field(curve) itself may not be related to RMT
buts rotational curves are. The RMT behavior
of gravity rotational curve supports that gravity behaves on a Quantum mechanical level.

Gravity and complex eigenvalues

The "Rotational curve correlation conjecture" is based on the idea that the totality of gravity rotational curves($1/r^2,r^2/r^2,r^2,-1/r^2,-r^2/r^2,-r^2$) and universe inflationary/deflationary curves ($1/x^2,(x^2/x^2),x^2,-1/x^2,(-x^2/x^2),-x^2$) obeys the normal distribution curve in the complex plane. Gravity & the universe on the large scale has a normal distribution curve in the complex plane.

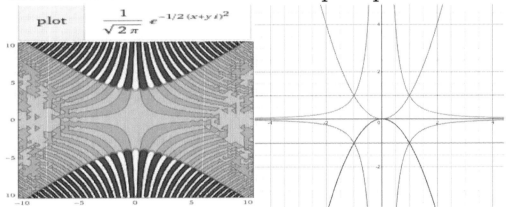

The Rotational curve correlation conjecture is based on Complex conjugate pairs. The decreasing curve is from real matter and increasing curve is from imaginary(FTL) matter and the flat curve is a overlap of them both. The mirror image is its complex conj. pair.

Essentially, we have a complex gravity rotational and inflation/deflationary cosmological system. If rotational curve and inflationary curves are complex in nature then what about gravity? Is gravity itself complex as well? The answer appears to be yes if we view <u>gravity as a Complex eigenvalue system</u>. Graphs below are based on complex eigenvalues. Our galaxy is like a spiral sink.

Spiral sink   Center   Spiral source

In the complex conjugate mirror universe we have mirror image pairs of the Spiral source, center and Spiral sink. Those 6 types of spiral systems corresponds to the 6 types of Rotational and inflationary/deflationary curves. Complex gravity normal distribution curve is based on a complex conj. pair big bang analogous to a complex conj. QM Ψ wave function. <u>Our universe could exist in a superposition of many states(multiverse) like a QM wavefunction exists in a superposition of many states</u>.

Complex normal distribution curves and critical points

Notice how the graph of the <u>complex nor-mal distribution curve</u>, <u>complex Riemann Xi function</u> and <u>6 rotational curves(graph on far right)</u> has a resemblance to critical points.

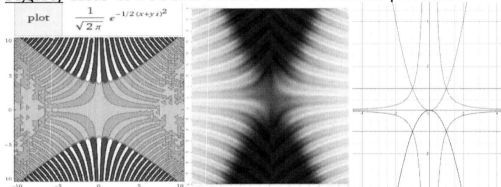

plot $\dfrac{1}{\sqrt{2\pi}}\, e^{-1/2\,(x+y\,i)^2}$

<u>Graph of critical points</u>

The graphs above have a saddle point pattern & the parabolas are sink or source nodes.

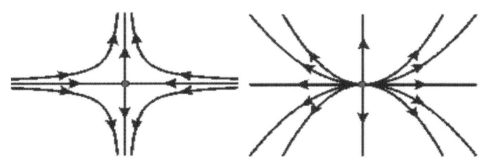

Saddle point   *Unstable node (source)*

In the context of galaxy rotational curves the spiral-in is our galaxy. The spiral-out would be related to the reversed gravity from the FTL space flow from inside a black hole. The interior of a black hole were space flows FTL is causing the external r^2 rotational ghost curve. The overlap of the real $1/r^2$ and imaginary ghost curve r^2 is the flat rotational curve $r^2/r^2=1$. That flat rotational curve is analogous to a centered system. We can also apply this model to our universe were we have a overlap between an inflationary x^2 imaginary energy) and deflationary curve ($1/x^2$ real energy). Their overlap can explain the uniform temperature of the universe and why the geometry of the universe is flat.

New correlation between QM and the cosmological

We thought that their was a connection between the nuclear strong force and gravity (Gravitons). Now we see that the correlation is between energy levels inside heavy atomic nuclei and the gravitational energy behavior of rotational curve systems and inflationary/deflationary universe curves.

Symmetry or no symmetry in Random matrix gravity(RMG)

For Quantum mechanics Random matrices we have 3 types: GUE, GSE and GOE. The 3 different types of Random matrices related to symmetry. If the "Rotational curve correlation" conjecture is true then the complex normal distribution curve of rotational curves of gravity (Random matrix of gravity) may fit a particular matrix symmetry analogous to GUE, GSE,GOE. For instance in this random matrix theory of gravity its suggesting Quantum gravity may not have time reversal symmetry like a GUE, or time symmetry like GSE or time symmetry like a GOE but no rotational symmetry. If the rotational curve correlation conjecture fits a random matrix then it supports a quantum nature to gravity where the random matrix represents a quantum randomness in gravity. Just as numbers are random in a Random matrix, it is quantum particles or energy fluctuations or discrete space-time itself that is of a Random nature & causes the complex rotational curves to behave like a Random matrix.

A flat rotational curve is analogous to eigenvalues and may consist of eigenvalues of gravity where the visible matter follows its layout. Also in the overlap of our expanding universe from imaginary energy and contracting universe from real energy there interacting creates eigenvalues related to flat space time. On a quantum scale eigenvalues of space time repel each other like Zeta zeros or energy levels in a heavy atomic nuclei. Gravity and its quantum nature may be related to Random matrix theory rather than Loop quantum gravity or string theory. What's interesting is that Random matrix gravity may have no time symmetry like a GUE or time symmetry like a GSE or time symmetry but no rotational symmetry like a GOE. In either case Random Matrix Gravity appears to point towards a single Gravity operator equation for Quantum Gravity analogous to a GUE, GSE or GOE in Quantum mechanics.

Computer simulation tests

Before we look for experimental verification we can first test the Random matrix gravity model through Computer simulations.

The Gravity unitary Ensemble (GRUE) Equation: Random Matrix Gravity (Highlight of the "Rotational curve correlation conjecture".)

When we compare the graphs below it indicates that the gravitational rotational curve systems or inflationary universe curves <u>can be expressed as a single equation</u> just like a GUE or GSE or GOE equation.

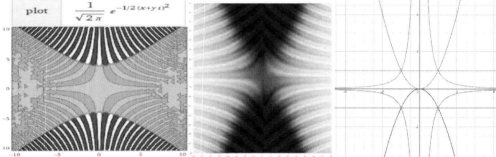

For example, below is the equation for the Gaussian Unitary Ensemble.

$$\frac{1}{Z_{GUE(n)}} e^{-\frac{n}{2} \operatorname{tr} H^2}$$

The GUE random matrix is captured in one equation. Of course their are supplemental equations that relate to the Random matrix such as the joint probability density for the eigenvalues but it's all based on the GUE equation. The 6 rotational curves can be described in a single "Gravity unitary Ensemble (GRUE)" equation. What's interesting is that the quantization of gravity starts with the Random variables. In other words some fundamental variables whether its gravity particles or discrete chunks of spacetime or something else are the random variables for the Gravity Unitary Ensemble(GRUE). The eigenvalues(energy levels) for this GRUE is represented by the flat rotational curves in the model. GRUE relates to String Theory and LQG(Loop Quantum Gravity) in the sense that both points towards a more fundamental structure of spacetime or particles. For GRUE the Random variables in the Gravity Unitary Ensemble(GRUE) represent the fundamental components of spacetime or particles or something. GRUE can be captured in a single equation just like a GUE, GSE or GOE is expressed as individual equations! GRUE points towards a Complex conjugate pair big bang universe analogous to a complex conjugate pair QM Ψ wave function & its superposition of many states(multiverse)!

35

Explaining Blackhole Entropy!!!!

At the <u>heart of the "Rotational curve correlation conjecture"</u> model is the idea that the FTL space-flow inside a black hole is causing the x^2 rotational ghost curve. Meaning as space falls faster than the speed of light the laws of gravitational physics reverse just like in Special Relativity were physics reverse into Tachyons for FTL travel. We have a Black hole Event horizon and White hole E. horizon.

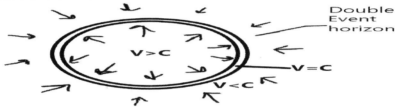

<u>The White hole inside a BH forces matter on the surface of the White hole like a Hologram.</u>
The 3 parts of a black hole: inward flow, Double Event horizon(equilibrium) and outward flow can be compared to 3 different rotational systems: Spiral in, centered and spiral out.

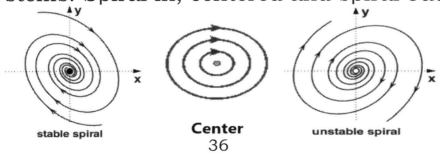

stable spiral   Center   unstable spiral

Double Event horizon equilibrium

Once you go underneath the Event horizon you will hit a brick wall of repulsive space flow geodesics. Also you can't escape gravity of the BH Event horizon. The result are two different Event horizons like Hot & cold water

Maximum black hole Entropy

The two overlapping Event horizons causes a Equilibrium like hot water next to cold water. The Event horizon of the B. hole and W. hole have V=c in common. <u>The maximum entropy is the equilibrium between the two different Event horizons</u> analogous to where temperatures are equal and heat flow is zero.

Interpreting the BH entropy equation

Conjecturally the c^3 term is related to the FTL White hole inside the BH & the Gravity constant is related to the Black hole Event Horizon.

$$S = \frac{\pi A k c^3}{2 h G}$$

What about Black hole mergers?

If the interior of a black hole is a white hole based on FTL space flow geodesics analogous to the FTL physics of Special relativity for V>C then how can black holes merge? Meaning the imaginary energy(mass) of the interior of the black hole is causing spaceflow to flow away from the center. It's a repulsive effect. At first glance you may say that the repulsive interior of a black hole undermines the new black hole model because how can black holes merge if their interiors are repulsive. On the contrary, Black hole merger supports the new model of a black hole and its connection to the "Rotational curve correlation conjecture". In terms of QM a black hole merger is analogous to protons merging in an atom under the Nuclear strong force. A computer simulation will be needed to test it. Conjecturally, the White hole inside a black hole is similar to the Pauli exclusion principle for a Black hole were it prevents a Singularity.

The inward gravity of the Black hole is like the Strong Nuclear force holding the internal white holes inside. Keep in mind that the white holes are not simply time reversed black hole. They are the result of space flowing FTL underneath the Event horizon just like a Tachyon come into being for FTL Special Relativity. The laws of geodesics and energy reverse for FTL space flow like a V>C traveling particles has the reverse physics of a Tachyon.

The Montgomery pair correlation conjecture relates the spacing of the nontrivial zeros to the energy levels inside heavy atomic nuclei. The Rotational curve correlation conjecture is related to the internal cluster of FTL spaceflow white holes inside a large black hole(black merger). In which the white holes inside a black hole are forced together under the crushing inward BH gravity analogous to the Strong force holding protons together.

Flat rotation curve   Pair correlation

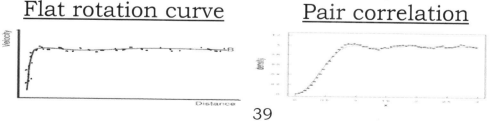

39

Hawking radiation vs. Whitebody radiation

The laws of physics for a White hole are reverse. At the center of the white hole may be imaginary mass analogous to a stationary Tachyon particle(imaginary mass star). It's Tachyon mass is pushing space away from it and have a W.H. Event horizon around it.
If Blackbody radiation exist in our universe then the white hole below the Event horizon has "Whitebody radiation". Meaning it's radiating energy in a White body radiation curve. The radiation from a Black hole isn't by Hawking radiation but by the White hole inside a black hole. The outward flow of space and energy from the White hole is hitting the Gravity Black hole Event horizon for a Equilibrium state and high entropy like placing cold water next to hot water.
The issue with Hawking radiation applied to the double Event horizon black hole model is that the white hole underneath is pushing space away. Conjecturally, nothing can go into the white hole Event horizon unless the virtual particle is traveling faster than light.

Meaning the virtual particle can't reach or interact with the center imaginary mass inside the white hole due to the repulsive space flow from it. If the black hole Event horizon splits virtual particles then the particle that goes beneath the Event horizon is hitting the White Hole Event horizon and never touches the central imaginary mass core.

A black hole is radiating but the radiation is coming from the White hole underneath the blackhole & not from Hawking radiation. The exotic physics of the imaginary mass(energy) of the white hole inside a black hole is causing mass reduction.

Complex big bang and Complex BHs

Our universe started with equal parts real energy(matter or $1/x^2$ curve) & imaginary energy(x^2 curve) and it's complex conjugate pair rather than a matter-antimatter pair. The universe is accelerating based on a X^2 curve because it's imaginary energy similar to Tachyon physics. It won't have a decreasing curve from the start but a increasing curve from the start. This complex universe gives raise to complex black hole structures.

Rotational curve correlation conjecture overview

This explore additional applications of the Montgomery pair correlation conjecture beyond relating the spacing of the nontrivial zeros to the spacing of energy levels inside heavy atomic nuclei. The conjecture is that the flat rotational curve galaxies is related to Random Matrix theory of Quantum gravity. Meaning something on a QM scale is random and is causing the correlation between graph of the Montgomery pair correlation conjecture to the Flat rotational curves for galaxies.

Flat rotation curve   Pair correlation

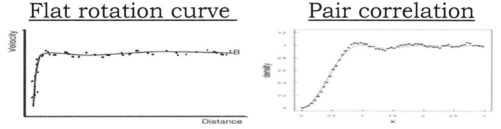

The 6 rotational curves of galaxies has correlations to the Riemann Xi function graph.

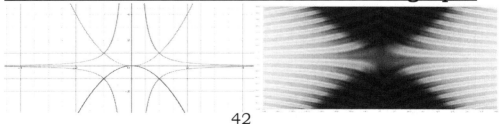

Additional gravity not MOND gravity

The Rotational curve correlation conjecture don't modify gravity at large scale but states that separate and additional gravity comes into place at large scale.
If we interpret the FTL spaceflow under the Event horizon in the context of FTL Special Relativity then the spaceflow underneath the Event horizon reverses and so does time. Effectively we have a white hole underneath a black hole. It's double Event horizon sandwhich. By having a white hole underneath a black hole we have two opposing rotational curves. The Gravity of the black hole has a external $1/\sqrt{r}$ rotational curve. However, the whitehole inside has a external $\sqrt{r}$ curve. The two curves overlap $\sqrt{r}/\sqrt{r} = 1$ to create a flat rotational curve. Notice the $\sqrt{r}$ curve from the white hole has more of a influence further away as Gravity declines.

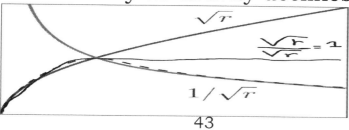

Dark Imaginary energy & the Big Crunch

If we had a complex big bang of equal parts real and energy energy(mass) then Imaginary energy (mass) can explain the expansion and acceleration of our universe. As the universe speeds up due to Imaginary energy its losing energy based on FTL tachyon physics. The universe expansion rate is approaching infinite speed as Energy approaches zero.

Will there a big rip to the universe?
On the surface it may appear the universe is headed for a big rip but keep in mind gravity from real mass is still present. There is a war between Tachyon energy level and gravitational energy level. Please note from the complex conjugate pair big bang our universe has all curves simultaneously. It's both expanding and contracting at the same time. The horizontal line is a overlap between the forces.

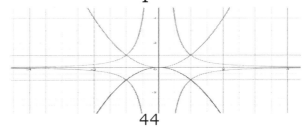

Heading towards a big crunch
(As tachyons lose energy as the universe speeds up it will reach a point were it's repulsive energy is less than the pull of gravity)

1. At the early universe the repulsive Tachyon energy(x^2 curve) is high(like a full tank of gas or like a heavyweight MMA fighter full of stamina) & it's repulsive strength overpowers contracting gravity($1/x^2$ curve).

2. However, tachyons are losing energy as the universe accelerate. It's like a MMA fighter gassing out. Their will be a point were the repulsive force of imaginary energy tachyons equals the attractive pull of gravity analogous to the equator on a sphere.

3. At some acceleration point in the future the Tachyons will have lost so much energy that it's repulsive energy is weaker then the gravitational pull from real mass. It's like a big MMA fighter who gassed out and stamina is gone. When the pull of gravity is stronger than the repulsive force of tachyons the universe will collapse towards a big crunch. Also the complex conjugate pair universe will be headed for a big crunch as well.

45

8. Universal L-functions & Physics Operators

Zeta and Xi have the same nontrivial zeros.

Z. function	Riemann Xi	Complex distri.

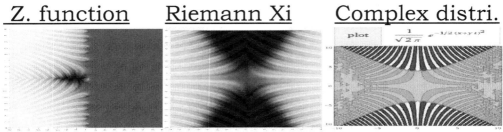

What's interesting is that the rotational curves of galaxies, universe inflationary-deflationary curves, QM harmonic oscillators, Complex GUE fit the behavior of a <u>complex normal distribution curve</u> like the Xi function. What if they have L-functions associated with them like the Riemann Xi-function have the Riemann Zeta function associated with it?

Gravity curves   QM oscillators   Complex GUE

$(+,- x^2)$ QM Oscillator $+,- x^2$

$(+,- 1/x^2)$ QM Osc. $+,- 1/x^2$ $\dfrac{1}{Z_{Gue}(s)} = \dfrac{1}{2^{1/2\,(x+yi)}\,\pi^{1/2\,(x+yi)^2}}$

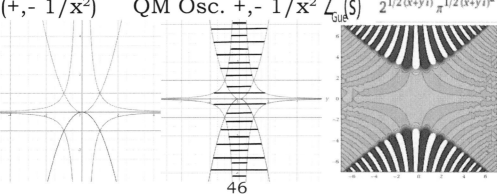

L-functions for Gravity curves, QM h. oscillators & Complex GUEs

The Riemann Xi function(similar to a complex normal distribution curve) is related to the Riemann Zeta function. What if the Gravity curves graph, QM harmonic oscillators graph & Complex GUE graph that's structured like the Xi function are related to L-functions? In which complex normal distribution curve physics systems can be reduced to a sum of some fundamental quantity in a infinite series analogous to a purely ∞ number series. The analytic continuation of this infinite sum physics system can be analytically continued to the rest of the complex plane. The complex normal distribution curves we see in gravity, QM harmonic oscillators, Complex GUEs behave like the Riemann Xi function. Just as the Xi function has a associated Riemann Zeta function it's possible complex normal distribution curves physics systems have an associated L-function form to them as well. The future of mathematical Physics may be Complex normal distribution curves physics systems & their associated L-functions!

9. <u>The Complex Prime Number conjecture</u> <u>(CPNC)</u>

What makes the <u>Riemann Zeta function,</u> <u>Riemann Xi function and the Riemann Hy-</u> <u>pothesis</u> so special is their connection to the distribution of Prime Numbers. This chapter explores PNT and the logarithmic integral li(x) in the complex plane like Bernhard Riemann explored the Euler Zeta function in the complex plane. In much the same way Riemann discovered Complex analysis the use of complex variables in PNT reveals a Complex Prime Number conjecture(CPNC).

When you study Number Theory you'll notice that prime counting functions only use real variables while L-functions use complex variables. In the proof of the Prime Number Theorem(PNT) it uses the Complex Zeta function to prove it. However, actual Prime counting functions are not studied in the field of Complex Analysis. This chapter explores using imaginary variables in Gauss PNT and li(x) to yield complex sum prime count estimates. What we find are breakthrough discoveries in the complex nature of PNT.

Complex Prime Number conjecture
(CPNC)
ix/log(ix) = P+Pi = P

It uses imaginary variables in Gauss original
Prime Number Theorem to give complex sums

25i/log(25i) = 6.27+ 3.06i =9
100i/log(100i)= 19.45 + 6.6i = 25
1000i/log(1000i)= 137 +31.3i = 168
(10^6)i/log((10^6)i)= 71458+ 8124i= 79582

If you add the real and imaginary part as
whole number their sum gives a better approximation than Gauss original x/log(x). The
exciting aspect is that we can now use imaginary variables to study complex prime counting sums. In which Gauss PNT can now be
included directly in the field of Complex Analysis. They are Prime counting functions with
imaginary or complex variables.

*Complex Prime Number Conjecture(CPNC) ex-
tends PNT into the complex plane like Bernhard
Riemann extended the Euler Zeta function to the
complex plane that spawned Complex analysis.*

49

The following gives a short list of values for
ix/log(ix) = Primes+Primes(i).

10i/In(10i) = $\underline{2}$.022+$\underline{2}$.964i = $\underline{4}$ and true value is 4

20i $\underline{2}$.746+$\underline{5}$.236i   = $\underline{7}$ and the true value is 8

25i $\underline{3}$.061+$\underline{6}$.273i =$\underline{9}$ prime numbers less than 25 and true value is 9

30i   3.357+ 7.27 i =10 and the true value is 10

40i   3.909 + 9.179i 12   and the true value is 12

50i   4.419 + 11.007i   =15 and the true value is 15

60i (4.900+12.77i) = 16 and the true value is 17

70i   5.359 + 14.495i = 19 and the true value is 19

80i   5.799+ 16.178i   = 21 and the true value is 22

90i   6.224 + 17.828i =23 and the true value is 24

95i   6.430699919444582+18.643165i = 24 and true value is 24

100i 6.635+19.452i =25 and the true value is 25

200i (10.28+34.69i) = 44 and true value is 46

300i 13.464 + 48.889i = 61 and the true value is 62

400i 16.377 + 62.468i = 78 and the true value is 78

500i 19.115 + 75.624i = 94 and the true value is 95

600i 21.722 + 88.46i =109 and the true value is 109

700i 24.228 + 101.043i = 125 and the true value is 125

800i 26.651 + 113.415i =139 and the true value is 139

850i (27.83+119.532i) = 147 and the true value is 146

900i 29.005 + 125.609i = 154 and the true value is 154

950i/ln(950i) = (30.159+131.64i) = 161 and the true value is 161

1000i 31.3+137.647i = 168 and the true value is 168

1500i/ln(1500i) = (42.11+196.062i) = 238 and the true value is 239

2000i 52.15+ 252.349i = 304 and the true value is 303

2500i (61.66+307.147i) = 368 and the true value is 367

3000i 70.789 + 360.813i = 430 and the true value is 430

3500i 79.607+413.57i) = 493 and the true value is 489

4000i   88.174 + 465.574i = 553 and the true value is 550

4500i   (96.53+516.9i) = 613 and the true value is 610

5000i 104.706 + 567.737i = 671 and the true value is 669

6000i 120.6 + 667.918i = 787 and the true value is 783

7000i 135.992 + 766.506i = 901 and the true value is 900

8000i 150.971 + 863.768i = 1013 and the true value is 1007

8500i = (158.32+911.96i) = 1070 and the true value is 1059

9000i 165.603 + 959.9i = 1124 and the true value is 1117

10,000i 179.935 + 1055.049i =1234 and the true value is 1229

25,000i (373.94+2410.733i) = 2784 and true value is 2762

50,000i (657.044+4525.77i) = 5182 and the true value is 5133

100,000i 1163.425+ 8527.155i= 9690 and the true value is 9592

100,500i (1168.24+8566.215i) = 9734 and the true value is 9632

101000i (1173.06+8605.267i) = 9778 and the true value is 9673

120000i = (1353.68+10078.764i) = 11432 and true value is 11301

125000i = (1400.46+10463.48i) = 11863 and true value is 11734

130000i = (1446.97+10847.04i) = 12294 and the true value is 12159

200,000i = (2074.26+16118.34i) = 18192 and true value is 17984

600,000i = (5251.098+44476.934i) = 49728 and true value is 49098

1,000,000(i)   8,124.708+71,458.651i = 79,582 and true value is 78,498

10 million(i) 59,894.528 + 614,583.64i = 674,477 and the true value is 664,579

100 million(i)   459,580,875+ 5,390,000i =5,849,580 and true value is 5,761,455

500 million(i) (1945629.40+24809828.68i) = 26755458 and true value is 26,355,867

1 Billion(i) = (3636766.41+47979280.31i)= 51616046 and true value 50,847,534

10 Billion(i) = 29489816+432282721i =461,772,537/ true value is 455,052,51

100 Billion(i) = 244,000,000+3,930,000,000i
= 4,174,000,000/true value is 4,118,054,813
500 Billion(i) =
(1078671083.29+18498327990.55i)
= 19,576,999,073 and true value is
19,308,136,142
1 Trillion(i)= (2050806632.03+3607462048
8.81i) = 38125427120 and the true value is
37,607,912,018
$(10^{15})i$ = 1.31 x 1012 +2.89 x 1013
=30,210,000,000,000/ true value is
29,844,570,422,669
$(10^{16})i$ = (1.155e13+2.70e14i) =
2.824935767660819e14 and the true value is
279,238,341,033,925
$((10^{20})i)/ln((10^{20})i)$ =
2.242930508256443e18 and the true value is
2,220,819,602,560,918,840
$((10^{22})i)/ln((10^{22})i)$=
2.033323650147539e20 and the true value is
201,467,286,689,315,906,290
$(10^{26})i$ = 1.713011104603802e24 and the
true value is 1,699,246,750,872,437,141,327
,603

(2*10^22)i/ ln((2*10^22)i) = 4.010228352640418e20 and the true value is 397,382,840,070,993,192,736
(4*10)i^22 /ln((4*10^22)i) = 7.910701891491991e20   and the true value is 783,964,159,847,056,303,858   It is a difference of 7.106029344933675e18 from the true value. This is a better approximation compared to the original PNT(x/log(x)) that use only real input variables.
((2^81)i)/ln((2^81)i) = (1.203895408972707e21+4.303079414491 387e22i) = 4.423468955388658e22 and the true value is 43860397052947409356492 were the difference(overshoot) is only 3.742925009391735e20 from the true value. The solution is based on adding the whole numbers from the real part & imaginary part of the complex approximation value. Compare that value to Gauss PNT where the approximation is 4.306447616692965e22. The difference(undershoot) of Gauss PNT from the true value is 7.959208860174621e20.

Oscillatory behavior of ix/log(ix)

The early values oscillate the true values.

Now let's see the graphing behavior of CPNC: Plot ix/log(ix) from -1000 to 1000.

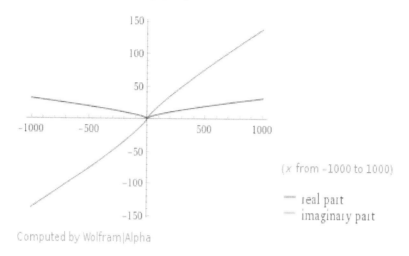

(x from –1000 to 1000)

real part
imaginary part

Computed by Wolfram|Alpha

Notice how the imaginary odd curve stops at around 160. Primes less than 1000 is 168.

Whats interesting about the graph is that it is taking the traditional PNT x/log(x) that uses real variables and exploring the function using imaginary variables. It suggests that Prime counting functions may work better using complex input variables rather than using only real input variables. We can also plot li(ix) ranging from a negative number to a positive number.

57

What's even more interesting is the 3D graph plot of ix/log(ix) as follows:
Plot 3d ix/log(ix) from -10^7 to 10^7. Graphs are made using Wolfram Alpha

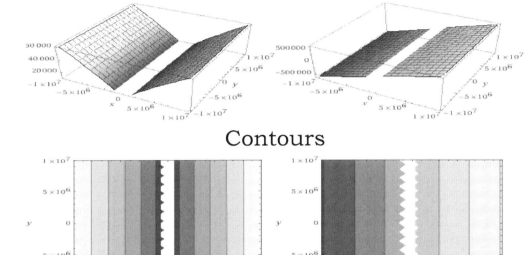

Contours

Trigonometric functions for ix/log(ix) or li(ix)

What's so amazing are the Trig. wave patterns. The vertical lines and Trig. waves down the middle may be due to a Trig. based logarithmic form of cosx + i sinx. We can get the same graph above with li(ix). Just as e^ix = cosx+ i sinx, the li(ix) = logarithmic version of cos x + i sin x Trig. waves for Prime counts.

58

The imaginary Logarithmic integral li(ix)

The estimates of ix/log(ix) gives superior results over Gauss original x/log(x). Naturally, the next step is to compute and graph li(x) as li(ix). However, it behaves differently. The Prime counts are only in the imaginary parts.
Li(10i) = 3.493 + 5.52i = 5 and their are 4 primes less than 10
Li(100i) = 12.09 + 25.68i = 25 and true value is 25
Li(50i)= 8.3.+15.412.=15 and true value is 15.
Li(1000i) = 47.966 + 163.73i = 163 and true value is 168
Li(10,000i) = 241.41 + 1195.7973i = 1195 and true value is 1229
Li(100,000i)= 1445.939 + 9402.06i = 9402 and true value is 9592
Li(1,000,000i) = 9652.729 + 77407i = 77407 and true value is 78,498

A oscillatory behavior

Notice how early values gives a Prime count more than the true value and afterwards others are exact values and then others are less than the true value. It's value oscillate.

More computation from li(ix)

Notice for the higher values the prime counts for li(ix) goes back to being more than the true value and then back below the true value in a oscillatory manner.

The Prime counts are in the imag. parts.

Li(10,000,000i) = 69112.205 + <u>657594.637i</u> = and true value is 664,579

Li(100,000,000i) = 519534.322 + <u>5714719i</u> and true is 5,761,455

Li(1,000,000,000i) = 404877 + <u>50523515i</u> = and the true value is 50,847,534

Li (10 Billion) = <u>455,968,140</u> and the true value is 455,052,511

Li(Trillion(i) = 221760552.2 + <u>37476483367i</u> and the true value is 37,607,912,018 where the difference from true value is 90,331,901

Li(10^{14}i)= 3.212923125908e12 and the true value is 3,204,941,750,802 and the difference is 7,981,375,106

A highly oscillatory Prime function

For ix/log(ix) =P+Pi and li(ix) the Prime counts gives an oscillatory. However, the li(ix) are more common. Ix/log(ix) oscillatory starts early but becomes very larger as we increase.

Oscillatory Complex Prime functions

The remarkable aspect of ix/log(ix) and li(ix) is that it does generate Prime counting solutions in complex form. In doing the totals oscillate the true values. For ix/log(ix)=P+Pi we see oscillation around the true value at the beginning but as we grow larger the oscillations are longer(huge). However, for li(ix) were the imaginary part total gives the prime count the oscillations around the true value are more frequent. We can also study the Riemann Prime counting function using imaginary variables. It will oscillate the true value.

$$J(x) = Li(x) - \sum_{p}^{\infty} Li(x^{\rho}) - \log 2 + \int_{x}^{\infty} \frac{1}{t(t^2 - 1)\log t} dt$$

Example. RiemannR(1000000i) = 9571..+ 77350i and the true value is 78,498. Without imaginary variables the Riemann Prime function already oscillated the true values because its related to Zeta zero. Using RiemannR(ix) would cause it to oscillate differently. The following section shows how RiemannR(ix) generates better prime estimates than RiemannR(x) at high count prime estimates!

Surpassing Riemann's Error term!

With traditional Prime counting functions:
$$x/\log(x)$$
$$\text{li}(x)$$
$$\text{RiemannR}(x)$$
the Riemann Prime counting function gives the best approximation for the number of Primes less than a given number.

$$J(x) = Li(x) - \sum_{p}^{\infty} Li(x^p) - \log 2 + \int_{x}^{\infty} \frac{1}{t(t^2 - 1)\log t} dt$$

For the complex Prime number conjecture(CPNC) we introduce imaginary variables.
$$ix/\log(ix)=P+Pi$$
$$\text{li}(ix)$$
$$\text{RiemannR}(ix)$$
The imaginary PNT ix/log(ix) provides better estimates than x/log(x). For li(ix) and RiemannR(ix) they don't(for low estimates) give better approximations than the traditional li(x) or RiemannR(x) counts. However, as we go larger in Prime estimates the new li(ix) and RiemannR(ix) actually gives better counts than li(x) and Riemann(x)!

When we compare the approximations of
li(x) and RimennR(x)
to their imaginary counterparts
li(ix) and RiemannR(ix)

It appears that li(ix) and RiemannR(ix) gives inferior results to li(x) and RiemannR(x). This is in contrast to ix/log(ix)=P+Pi better results than x/log(x). However, if we go larger in approximations for li(ix) and RiemannR(ix) they give a better approximation than li(x) and RiemannR(x). The following compares low and high values between the different Prime counting functions:

$25/\log(25)= 7.7...$
$25i/\log(25i)= \underline{3}.06..+\underline{6}.27i..= 9$
True value less than 25 is 9

$li(25)=11.5$
$li(25i)=5.77..+ \underline{9}.57i.. = 9$
True values less than 25 is 9

$RiemanR(25)= 8.87..$
$RiemannR(25i)=3.3..+\underline{7}.82i..= 7$
True value less than 25 is 9

$$1000/\log(1000)= 144$$
$$1000i/\log(1000i)= \underline{31}+\underline{147i} = 168$$
True value less than 1000 is 168

$$\text{li}(1000)=177$$
$$\text{li}(1000i)=47..+ \underline{163i}.. = 163$$
True values less than 1000 is 168

$$\text{RiemanR}(1000)= 168$$
$$\text{RiemannR}(1000i)=39..+ \underline{159i}..=159$$
True value less than 1000 is 168

$$1000000/\log(1000000)= 72382$$
$$1000000i/\log(1000000i)= \underline{8124}+\underline{71458i}= 79582$$
True value less than 1000000 is 78498

$$\text{li}(1000000)=78627$$
$$\text{li}(1000000i)=9652.+ \underline{77407}..i$$
True values less than 1000000 is 78498
$$\text{RiemanR}(1000000)= 78527$$
$$\text{RiemannR}(1000000i)=9571+\underline{77350i}$$
True value less than 1000000 is 78498
We see at this range, li(x) & RiemannR(x) is better than li(ix) and RiemannR(ix).

Surprising math approximations

At low values, li(x) and RiemannR(x) gives better approximations but as we go higher in approximations something remarkable happens for li(ix) and RiemannR(ix). The approximations of li(ix) and RiemannR(ix) are closer than li(x) and RiemannR(x) to the true values!

$$li(10^{15}) = 2.984..10^{13}$$

$$RiemannR(10^{15}) = 2.9844 \times 10^{13}$$

$$li((10^{15})i) = 1.39 \times 10^{12} + \underline{2.977i \times 10^{13}}$$

$$RiemannR((10^{15})i) = 1.39 \times 10^{12} + \underline{2.977i} \times 10^{13}$$

$\underline{\text{True values less than}} \quad \underline{10^{15} =}$
$\underline{2.8953 \times 10^{13} \text{ (best known value)}}$

$$li(10^{18}) = 2.473..10^{16}$$

$$RiemannR(10^{18}) = 2.473 \times 10^{16}$$

$$li((10^{18})i) = 9.606 \times 10^{14} + \underline{2.4702i \times 10^{16}}$$

$$RiemannR((10^{18})i) = 9.606 \times 10^{14} + \underline{2.4702i}$$
$$\underline{\times 10^{16}}$$

$\underline{\text{True values less than}} \quad \underline{10^{18} =}$
$\underline{2.412 \times 10^{16} \text{ (best known value)}}$

At high values the approx. of li(ix) & RiemannR(ix) get closer to the true value than li(x) and RiemannR(x). Its amazing because we thought nothing could get better than Ri(x)!

li(10^39)= 1.126..10^37
RiemannR(10^39)=1.126 x 10^37
li((10^39)i)= 1.1992x10^35+1.125i x 10^37
RiemannR((10^39)i)=1.992 x10^35+1.125i
x10^37
True values less than   10^37 = 1.113
x10^37 (best known value)

The difference between li(x),Ri(x) and li(ix)
,Ri(ix) at high prime estimates are not huge
but the amazing discovery is that the solutions of the imaginary parts of li(ix) and Rie-
mannR(ix) gives closer Prime counting estimates to the true value than li(x) and R(x).

Riemann's Prime counting function has been dethroned

We see the Complex Prime Number conjec-
ture(CPNC) gives better estimates: ix/log(ix)
is better than x/log(x) and li(ix),RiemannR(ix)
gives better estimates(as the number get big-
ger) than li(x) and Riemann(x). It still has be
proven but ix/log(ix), li(ix),Ri(ix) are better
prime counters. I don't know exactly when
li(ix) & Ri(ix) become better but it's between
1 million < 15 million prime count range.

The highlight of the Complex Prime Number Conjecture(CPNC)

It's amazing how we can achieve better Prime counting functions than RiemannR(x) using RiemannR(ix). However, the highlight of CPNC is in the <u>Complex Prime Gap function</u>. For instance, let's take the average Prime gap function log(x). It counts the average spacing between Primes. Now lets use an imaginary variable. If we compute log(1000i) we get 6.9+1.57i. The conjecture is that the real part "6.9" is giving the average "even Prime gap" and "1.57i" is giving the one and only odd Prime gap spacing between 2,3. That is why regardless of how large you go such as log(1,000,000,000i) you always have 1.57i as the imaginary part. The function log(ix) is giving the average even and odd Prime gap spacing. It's amazing because the even and odd Prime counts are capture in the complex expression of log(ix)= Even+ odd(i) Prime gaps.

The Power series of Prime Gaps

When we look at e^ix it's represented by the series below. Because e^ix and log(ix) have inverse growth properties, the function log(ix) should have a similar power series to e^ix. The function log(ix) represent the average even and odd spacing Prime number spacing.

$$e^{ix} = 1 + ix + \frac{(ix)^2}{2!} + \frac{(ix)^3}{3!} + \frac{(ix)^4}{4!} + \frac{(ix)^5}{5!} + \frac{(ix)^6}{6!} + \frac{(ix)^7}{7!} + \frac{(ix)^8}{8!} + \cdots$$

$$= 1 + ix - \frac{x^2}{2!} - \frac{ix^3}{3!} + \frac{x^4}{4!} + \frac{ix^5}{5!} - \frac{x^6}{6!} - \frac{ix^7}{7!} + \frac{x^8}{8!} + \cdots$$

$$= \left(1 - \frac{x^2}{2!} + \frac{x^4}{4!} - \frac{x^6}{6!} + \frac{x^8}{8!} - \cdots\right) + i\left(x - \frac{x^3}{3!} + \frac{x^5}{5!} - \frac{x^7}{7!} + \cdots\right)$$

$$= \cos x + i \sin x.$$

For example, log(100i)= 4.6+1.57i were 4.6 represent the average even gap Prime spacing and 1.57i represent the one and only odd Prime gap between 2,3. If log(ix) has a power series like e^ix = cos x + i sin x then log(ix) should have a similar power series based on it's version of cos x+ i sin x.

68

The conjecture is that the even Prime Gaps(2,4,6,8,10...) on the number line is directly related to the even number power series within the cosine(x) part of log(ix). Also the odd Prime gap between 2,3 is related to the i sine(x) power series of log(ix). The power series of log(ix) is insight into the even and odd Prime gap function on the number line. What's amazing about this conjecture is that it captures Twin Primes, sexy Primes (Prime numbers that differ by three), cousin Primes-(Primes that differ by 4) and so on. The idea is that the power series of log(ix) is based on those groups of Primes.

The even gaps(2,4,6,8..) + Odd Prime (2,3) is analogous to the cos x + i sin x form for log(ix). For i sin x its always 1.57i because there is only one odd Prime between 2,3. Please note the power series for log(ix) is different from e^ix= cos x + i sin x. I am just using the power series of e^ix as a example. Once the power series for log(ix) is determined then it should give the template for the Prime gap. It will capture how every even and odd Prime gap spacing behaves.

The complex nature of Prime gaps

The equation

log(ix) = even prime gaps + odd Prime gap

Example,   Log(100i)= 4.6+1.57i

shows us the complex nature of the distribution of Prime Numbers. In which the even
Prime gap distribution such as 2,4,6,8,10...
can be described by a cosine function and
the only odd prime gap between 2,3 is described by the sine function and the value
1.57i. What's amazing is that the equation
log(ix) captures both the Even and Odd prime
gaps. It supports the idea that we can learn
even more about the distribution of Prime by
extending the PNT into the CPNC(Complex
Prime Number conjecture).

Euler formula in Prime Gaps
The Even space Prime gaps and the odd
Prime gap between 2,3 perfectly relates to the
Euler's formula e^ix = cos x + i sin x.

The amazing property of having the equation

log(ix)= even prime gaps+ odd gap

Example, Log(100i)=4.6+1.57i

is that we have insight into the inverse of
log(ix) in the form of Euler's formula e^ix.

$$e^{ix} = 1 + ix - \frac{x^2}{2!} - i\frac{x^3}{3!} + \frac{x^4}{4!} + i\frac{x^5}{5!} - \frac{x^6}{6!} - i\frac{x^7}{7!} + \cdots$$

$$e^{ix} = \left(1 - \frac{x^2}{2!} + \frac{x^4}{4!} - \frac{x^6}{6!} + \cdots\right) + \left(ix - i\frac{x^3}{3!} + i\frac{x^5}{5!} - i\frac{x^7}{7!} + \cdots\right)$$

$$e^{ix} = \left(1 - \frac{x^2}{2!} + \frac{x^4}{4!} - \frac{x^6}{6!} + \cdots\right) + i\left(x - \frac{x^3}{3!} + \frac{x^5}{5!} - \frac{x^7}{7!} + \cdots\right)$$

What's so exciting is that the even Prime
Gaps of 2,4,6,8,10, etc. are based on the
power series within log(ix). The Prime Gaps
are real and the only odd Prime Gap between
2,3 is imaginary like cos(x)+sinx(i).

71

A new Prime Gap Dirichlet series

For example, let's say we start with the Euler formula and create a Dirichlet series. We start with the Power series of e^ix = cos x + i sin x and raise it to a complex power.

$$= 1 + ix - \frac{x^2}{2!^s} - \frac{ix^3}{3!^s} + \frac{x^4}{4!^s} + \frac{ix^5}{5!^s} - \frac{x^6}{6!^s} - \frac{ix^7}{7!^s} + \frac{x^8}{8!^s} + \cdots$$

The power series of log(ix)= cos x + i sin x is similar to the series above just as a Dirichlet series include a special character L(x,s). Next, we want to determine if it has a Euler Product. What we want to achieve is to translate the Prime Gap properties of log(ix) = even Prime gaps + odd Prime gap into the language of L-functions and nontrivial zeros. When s=1 it will equal the Prime Gap function log(ix). This "Prime Gap L-function" could serve as a new tool to help prove conjectures about Prime Gaps.

The complex nature of Prime gaps

Just as the sin and cosine in the complex plane represent even and odd series the even and odd Prime gaps corresponds to behavior in the complex plane as $\log(ix) = x + 1.57i$.

Even Prime gaps+1.57i(Odd gap between(2,3))
$$\text{Cos}(x) + \text{Sin}x(i)$$

Even Prime(2)(cos)+ (Odd Primes(sin(i))

<u>On a fundamental level the behavior of Prime gaps are complex in nature.</u> For the study of Prime gaps such as proving the "Twin Prime" conjecture it possible complex number mathematics can be applied to prove it as true. The Complex Prime Number Conjecture (CPNC) starts off with oscillatory complex Prime counting functions such as $ix/\log(ix)$ and $li(ix)$ but the highlight is the Complex Prime gap conjecture as follows:

The even and odd Prime gaps are related to the Cos(x)+ sinx(i) function in the complex plane.

73

Approximation of the nth Prime using imaginary input variables

The Complex Prime Number conjecture state

$$\pi(x) \sim ix/\log(ix) = P+Pi= P$$

As a consequence, the approximation formula for locating the nth prime is as follows:

$$P_n \sim (ix)\log(ix)$$

Lets compute this formula and compare it to the actual location of the nth Prime. Just as with CPNC only add the whole positive real numbers from the real and imaginary parts.

$(10i)\log(10i) = -15.707+23.025i$ and the actual nth Prime is 29

$(100i)\log(100i) = -157.07+460.51i$ and the actual nth Prime is 541

$(1000i)\log(1000i)= -1570+ 6907i$ and the nth Prime is 7919

$(1345i)\log(1345i)= -2112+9689i$ and the nth Prime is 11093

If we add the real & imag. parts then it gives a slightly closer approximation to the nth Prime

The exciting aspect of using imaginary variable is that it works to some degree. Meaning it gives a slightly better approximation than just using real input variables such as xlogx. Lets exploring it's graphing behavior.
Plot (ix)log(ix) from -25 to 25
We want to find the nth Prime either from 0 to -25 or 0 to 25. Graph made by Wolfram Alpha.

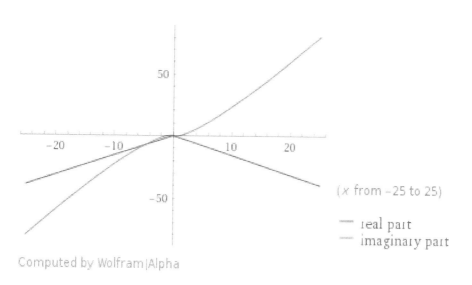

Computed by Wolfram|Alpha

The actual 25th Prime is 97 and notice where the odd curve stops at. It stops around in the -/+ 80 range. The function (ix)log(ix) is not perfect but it gives us new tools to study Prime functions using complex variables.

75

More observations when studying the behavior of ix/log(ix)

Plot the complex sums on a graph and they form a straight line. For , example, take the complex solutions such as 3+6i from 25i/log(25i) as x,yi plots. You'll see that they all fall on a straight line.

10/log10i= 2.021+2.9636 = 4 primes

25i/log25i = 3.061+6.273i =9 prime numbers less than 25 and true value is 9

100i/log100i = 6.635+19.452i =25 and the true value is 25

The following are some of the plots of the complex sums to show how their falling on a straight line. The same applies to li(ix) graphs

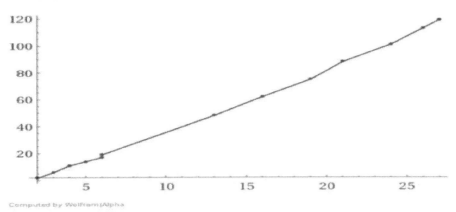

{{2, 2}, {3, 6}, {4, 11}, {5, 14}, {6, 17}, {6, 19}, {13, 48}, {16, 62}, {19, 75}, {21, 88}, {24, 101}, {26, 113}, {27, 119}}

Computed by Wolfram|Alpha

The proportion of the real group primes to the imaginary group primes is governed by the Natural Logarithm function

From the solutions of ix/In(ix) = P +Pi you notice that the real part is always smaller than the imaginary part. If you divide the imaginary and real parts together an amazing pattern emergences. For example, divide the number real part into the imaginary part as follows. Based on ix/In(ix) = P +Pi the imaginary value/real value = # are repeating in increasing order. In other words, as you keep dividing the i/r of the complex solutions of CPNC(Complex Prime Number conjecture) the final product will repeat in increasing order. For example, for 100i/log100i= 6.635+19.45 the i/r solutions of CPNT will be 19/5. Also 2i/log(2i)= 1.065+0470i will be 1/0.   If we do that for all complex solutions of ix/log(ix)= P+Pi then the value will grow slow were it corresponds to logarithmic growth.

Imaginary#/real#

2i/In(2i) = 1.07 + 0.47i   i/r   0/1=0
4i/In(4i) =1.43 + 1.26i   i/r   1/1 =1
10i/In(10i) = 2.02 + 2.96i   i/r   2/2 = 1
25i/In(25i) = 3.06+6.27i   i/r   6/3 = 2
100i/In(100i) =   6.635+19.452i   i/r
19/6 = 3
600i/In(600i) = 21.72 + 88.46i   i/r 88/22 =
4
1000i/In(1000i) = 31.3+137.65i   i/r
137/31 = 4
7000i/In(7000i)= 135.992 + 766.506i   i/r
766/135 = 5
1,000,000(i)   8,124.708+71,458i   i/r
71,458/8124 = 8
1 Billion(i) = 3640000 + 48,000,000i   i/r
=   13
1 Trillion(i)= 2,050,000,000+36,100,000,000i
i/r = 17
(1016)i = $1.16 \times 1013 + 2.71i \times 1014$ i/r = 23
(1020)i = $7.4 \times 1016 + 2.17 \times 1018$ i/r = 29

(1040)i= (1.851152573667031e36+1.0854204
96158272e38i) . i/r = 58

(1060)i = (8.228674060964752e55+7.237305
781028176e57i) . i/r =88

(10154)i =((10^145)i)/ln((10^145)i)=
(1.40910346330897e140+2.99506806321390
63e142i)   i/r = 212

For the complex solutions of ix/log(ix) = P+Pi
that gives Prime number estimates their real
and imaginary values are not random. From
the previous list shown it appears the proportion of the two is following the natural loga-
rithm.
Other Complex Prime functions
For the overall Complex Prime Number conjecture were not restricted to simple complex
functions such as ix/log(ix) or li(ix), etc but
more sophisticated functions such as Cheby-
shev ψ(x) can use imaginary variables to compute complex Prime sums. We can study it's
3D trigonometric behavior at the Prime jumps
to learn more info about this function.

The Complex Prime density

Studying various behavioral properties of CPNC(Complex Prime Number conjecture) will be interesting. We can explore the complex density make up of the complex sums from $1/\log(ix)$. For example, we can compute

$1/\log(1000i)= 0.13+ 0.03i$
$1/\log(10000i)= 0.15+ 0.01i$
$1/\log(100000i)= 0.085+ 0.011i$
$1/\log(1000,000i) = 0.07+ 0.008i$

Graphs. plot $1/\log(ix)$ from -100 to 100 and from -10^6 to 10^6. Notice how the density of the imag. part is smaller than the real part. Understanding why could be interesting.

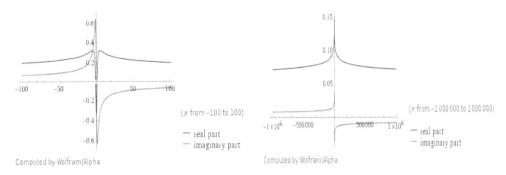

Exploring Guass PNT and other prime counting function with imaginary variables evolves PNT into the realm of Complex analysis were we'll see a deeper link between PNT & Zeta.

Complex proportion of primes

The Complex proportion gives the number of real part primes and imaginary part primes in the complex solutions of CPNC $ix/\log(ix)$ = $p+pi$ or complex $li(ix)$. The following is an example on how to determine the complex proportion of primes.

$25i/\log(25i) = 6+ 19i = 25$

Rather than saying 0.25 proportion of primes less than 100 in the complex solution I can be stated as

$0.06+ 0.19i = 0.25$

The complex proportion of primes value is telling us the Proportion of the real and imaginary part less than 100 as

$0.06 + 0.19i$

Exploring other Prime properties

We can also explore complex variables representing the $x+yi$ and $x+yi+zi$ plane.

e.g. $(25+25i)/\log(25+25i)$

e.g. $(25+25i+25i)/\log(25+25i+25i)$

The patterns from it's real and imaginary part totals can be and graphed like we have done for $ix/\log(ix)=P+Pi$.

Becoming the CPNT(Complex Prime Number Theorem) from the CPNC

The Complex Prime Number Conjecture(CPNC) is based on ix/log(ix) or li(ix). Because ix/log(ix) is so similar to the original PNT we should be able to prove CPNC using the Zeta function like PNT was proven using the Zeta function.

The following expression of the Complex Prime number conjecture (CPNC) shows as xi goes to infinity the expression goes to a complex 1 and the imaginary part to 0. As the real part is approaching 1, the imaginary value is approaching zero. Ultimately it will become just "1+0i". The following are just a few example of the asymptotic behavior of ix/log(ix)

(1229i)/((10000i/ln(10000i)))= (1.131950831
715873+0.1930508685630928i)
(9592i)/((100000i)/ln((100000i)))= (1.104319
810599944+0.1506707836661664i)

(78498i)/((1000000i)/ln(1000000i))= (1.0844
89947779079+0.1233043700607458i)
(783964159847056303858i)/((4e22i)/
ln(4e22i))= (1.01999943070089+0.030786200
56566507i)
43860397052947409356492i/
((2.417851639229258e24i)/
ln(2.417851639229258e24i))= (1.018482075
177986+0.02849461458458117i)

<u>Conventional or new method proof for CPNC</u>
The Complex Prime Number Conjecture
(CPNC) is very similar to PNT. The Zeta function was used to prove PNT. It may be possible to also use the Zeta function to prove
CPNC(Complex Prime Number Conjecture).
However, that remains to be seen. A proof
of the Complex Prime Number Conjec-
ture(CPNC) will cause it to become the Complex Prime Number theorem. Once Proven,
CPNT will become a part of the field of Complex Analysis. Meaning Gauss PNT using
imaginary or complex variables can be taught
in a Complex Analysis course rather than in
a separate area of Math(Number Theory).

Generating L-function type graphs from the Complex Prime Number Conjecture(CPNC)

Let's extend ix/log(ix) to complex variables such as (x+yi)/(log(x+yi)). Next let's add an exponential version of the function to it as follows. In which it's the <u>relatively inverse mirror image</u> of the CPNC. Notice how it stops on a line at around 0.5 like the Zeta function.

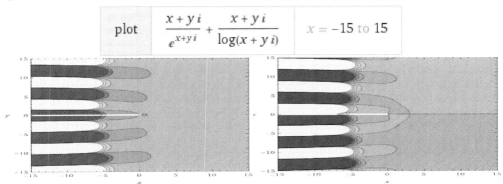

Now let's take it's inverse & notice how we see points aligning on a critical line approx. 0.5.

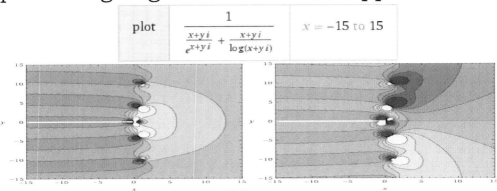

We can do the same with the Logarithmic integral using complex variables: li(x+yi).

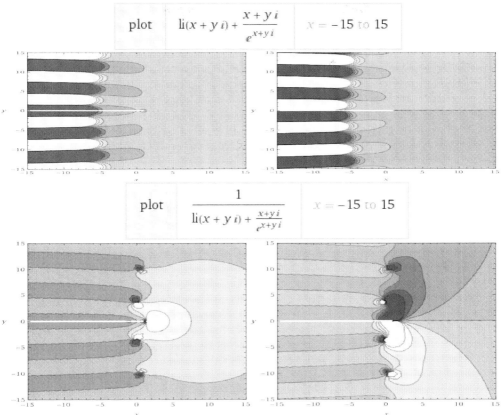

The point in showing you these graphs is that using complex variables(ix/log(ix)) or li(x-+yi) into CPNC generate Zeta like properties. By adding or subtracting a complex exp version to it's complex log version it generates a graph similar to a L-function that has points that align on a critical line(approx. 0.5 x-axis)

10. <u>Logarithmic analytic continuation of the Riemann Zeta function</u>

The point of this chapter is to provide a Number theory approach to prove RH. Bernhard Riemann extended the Zeta function using the Gamma function. His function reveals new discoveries on <u>Prime Numbers</u>. His extension worked but the complexity of the equations obscures the basic complex mapping behavior of the nontrivial zeros to the critical line. Meaning it doesn't explain in laymen terms how and why the nontrivial zeros map to the critical line. The following gives a different way to extend the Euler Zeta function by using the log function instead of the Gamma function for new insight. Let's invert the Riemann Zeta function based on the following relationship. $e^x = y$ means $log_e y = x$

$$\sum_{n=1}^{\infty} \frac{1}{n^s} = \frac{1}{1^s} + \frac{1}{2^s} + \frac{1}{3^s} + \cdots means \sum \frac{1}{Log_n\left(\frac{1}{1^s} + \frac{1}{2^s} + \frac{1}{3^s} + \cdots\right)} = s$$

$log_1(\text{Zeta(s)}) = 0$ in the infinite series.

What's interesting is that their exist a linear
scale between the logarithmic and exponential function as see in the following graph.

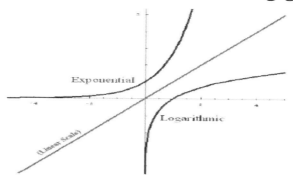

Just as e^x =y and log(y)=x has a linear scale
between them y=x so does the exponential
and logarithmic Zeta function below has a
linear scale line between them.

$$\sum_{n=1}^{\infty} \frac{1}{n^s} = \frac{1}{1^s} + \frac{1}{2^s} + \frac{1}{3^s} + \cdots means \sum \frac{1}{Log_n\left(\frac{1}{1^s} + \frac{1}{2^s} + \frac{1}{3^s} + \cdots\right)} = s$$

The take away is that the angles associated
with the exponential Zeta is reflected in the
logarithmic Zeta with respect to the linear
scale line between them. Conjecturally, the
linear scale line is a critical line between the
exponential and logarithmic version of the
Riemann Zeta function.

We can extend the exponential Zeta function by adding on the logarithmic Zeta function. Together they describe the together reflected growth. Now we have a new expression of the Zeta function based on Logarithmic analytic continuation. We can factor it to have a Euler Product assuming $Log_1(Zeta(s))=0$.

<u>A NEW ZETA-LIKE L-FUNCTION</u>

$$\sum_{n=1}^{\infty} \frac{1}{n^s + Log_n\left(\frac{1}{1^s} + \frac{1}{2^s} + \frac{1}{3^s} + \cdots\right)} = \prod_{p} \frac{1}{1 - \frac{1}{p^s + Log_p\left(\frac{1}{1^s} + \frac{1}{2^s} + \frac{1}{3^s} + \cdots\right)}}$$

What makes this logarithmic extended Zeta function so exciting is that the sum of the logarithmic and exponential Zeta function causes the Zeta function to be well behave in the entire complex plane except at 1(singularity). To <u>compute the nontrivial zeros of the Zeta-like function</u> we can use an alternating series <u>like the Dirichlet ETA function</u>.

$$\sum_{n=1}^{\infty} \frac{(-1)^{n-1}}{n^s + log_n\left(\frac{1}{1^s} + \frac{1}{2^s} + \frac{1}{3^s} + \cdots\right)} = \frac{1}{1^s + log_1(\zeta(s))} - \frac{1}{2^s + log_2(\zeta(s))} + \frac{1}{3^s + log_3(\zeta(s))} - \cdots$$

$$\sum_{n=1}^{\infty} \frac{1}{n^s + log_n(\zeta(s))} - \sum_{n=1}^{\infty} \frac{(-1)^{n-1}}{n^s + log_n(\zeta(s))}$$

$$\sum_{n=1}^{\infty} \left(\frac{1}{n^s + log_n(\zeta(s))} + \frac{(-1)^n}{n^s + log_n(\zeta(s))} \right)$$

$$\frac{2}{(2^s + log_2(\zeta(s)))} \sum_{n=1}^{\infty} \frac{1}{n^s + log_n(\zeta(s))}$$

$$\sum_{n=1}^{\infty} \frac{(-1)^{n-1}}{n^s + log_n(\frac{1}{1^s} + \frac{1}{2^s} + \frac{1}{3^s} + \cdots)} = (1 - \frac{2}{(2^s + log_2(\zeta(s)))}) \sum_{n=1}^{\infty} \frac{1}{n^s + log_n(\zeta(s))}$$

An example of log analytic cont. can be seen in the 1/(e^s + log(s)) graph below(right) were its graph has a critical line that relates e^s to log_e(s) like Zeta's critical line that relates s to 1-s. The same applies to $\Sigma 1/(e^s + log_e(s))$ or $\Sigma 1/(n^s + log_n(\zeta(s)))$ were Log_1(Zeta(s))=0. <u>This new form can be used to help prove a GRH.</u>

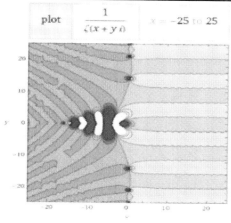

plot $\dfrac{1}{\zeta(x + y i)}$ x = −25 to 25

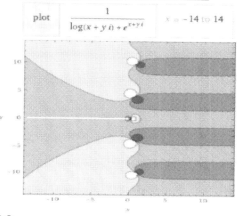

plot $\dfrac{1}{log(x + y i) + e^{x + y i}}$ x = −14 to 14

Mock L-function

It's not a infinite sum and don't have a Euler Product but it's graph is virtually identical to Zeta(left). Notice how both have a "mountain" at the center. Also notice the critical line for the Zeta like function is almost at 0.5 for its real part. Its a Mock L-function(right graph)

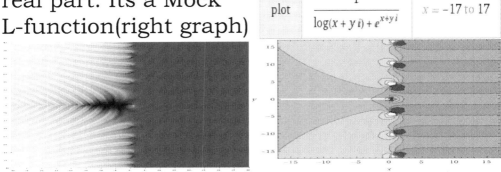

The inverse relationship between e^s and log(s) and the critical line between the two sides is perfectly analogous to the analytic continuation of Zeta of s to 1-s were it has a critical line between them at 0.5 on the x-axis. Log analytically is a new way to extend.

A new Zeta-like function

Graph comparison <u>above</u> supports the Zeta-like equation below assuming $\text{Log}_1(\text{Zeta}(s))=0$.

$$\sum_{n=1}^{\infty} \frac{1}{n^s + Log_n\left(\frac{1}{1^s} + \frac{1}{2^s} + \frac{1}{3^s} + \cdots\right)} = \prod_{p} \frac{1}{1 - \frac{1}{p^s + Log_p\left(\frac{1}{1^s} + \frac{1}{2^s} + \frac{1}{3^s} + \cdots\right)}}$$

90

Studying values of Mock L-functions and the Zeta-like function

One of the mysterious of the Riemann Zeta function beside RH and it's connection to Primes is the Odd positive integer values of Zeta. Even positive integers of Zeta have closed forms but it's a mystery as to whether the positive odd values have closed forms also.

Rather than attacking the Zeta function directly, we can study Mock L-functions that look and behave like Zeta. Possibility the values from the Mock L-function and Zeta-like function at odd positive integers can shed new light on the mysterious behavior of Zeta positive odd integers: Zeta(3),5,7. <u>Mock L-functions are a striped down Zeta function.</u>

$$1/(e{^\wedge}s + \log_e(s))$$
$$\Sigma 1/(e^s + \log_e(s))$$
$$\Sigma 1/(n^s + \log_n(\zeta(s)))$$ were $\text{Log}_1(\text{Zeta}(s))=0$.
Assuming $\text{Log}_1(\text{Zeta}(s))=0$

$$\sum_{n=1}^{\infty} \frac{1}{n^s + Log_n\left(\frac{1}{1^s} + \frac{1}{2^s} + \frac{1}{3^s} + \cdots\right)} = \prod_{p} \frac{1}{1 - \frac{1}{p^s + Log_p\left(\frac{1}{1^s} + \frac{1}{2^s} + \frac{1}{3^s} + \cdots\right)}}$$

Possible connection to the Euler–Mascheroni constant

Let's take a Mock L-function over the reals and do a converge test. To give context remember the Mock L-functions looks similar to the Riemann Zeta function in the complex plane.

Zeta graph

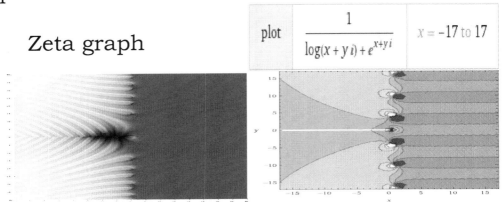

plot	$\dfrac{1}{\log(x+yi)+e^{x+yi}}$	$x = -17$ to 17

If we take the Mock L-function equation over the reals it converges towards a value.

$$\sum_{x=0}^{\infty} \frac{1}{e^x + \log(x)} \approx 0.567249$$

Partial sums:

What's interesting is that it's value of 0.567249.... looks close to the Euler–Mascheroni constant of
0.577215664901532...
If Σ(1/(e^x + log(x))) converges towards 0.567249.... We can also subtract them & it converges.

$$\sum_{x=0}^{\infty} \frac{1}{e^x - \log(x)} \approx 0.599431$$

It's graph in the complex plane will still resemble the Zeta function.

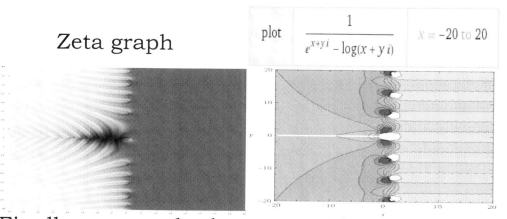

Zeta graph   plot $\dfrac{1}{e^{x+yi} - \log(x + y\,i)}$ $x = -20$ to 20

Finally, we can do the same with the Zeta-like function Σ(1/(n^x+log_n(x)) were It should con-verge towards a value. We can also try sub-traction such as Σ(1/(n^x-log_n(x)) assuming log_1(x)=0 for a finite value & see how it relates to the Euler–Mascheroni constant(0.567...).

Complex graph of the Euler–Mascheroni constant

The Euler-Mascheroni constant is defined as Zeta(1)-$\log_e$(n)=0.5772156649

On the previous pages I showed how similar equations exist and give similar values. Also when the equation was graphed in the complex plane it produced a graph that looked similar to the Zeta function. For example,

Zeta graph

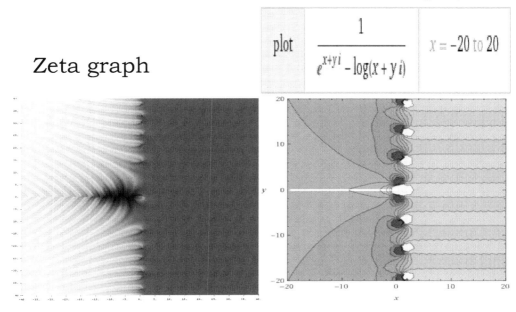

$$\text{plot} \quad \frac{1}{e^{x+yi} - \log(x+yi)} \quad x = -20 \text{ to } 20$$

Notice how when we subtract a complex log from a complex exponential function it's graph will resemble the Zeta function.

Likewise, If we take the Euler-Mascheroni constant and extend it into the complex plane then it's graph will resemble that of the entire Riemann Zeta function. Graph as a exercise.

$$\Sigma(1/n^s)-\log_e(s)$$

The expression above is extending the Euler-Mascheroni constant into the complex plane. Conjecturally, the graph of the complex function will resemble the Riemann Zeta function. Meaning it's graph will literally resemble the complex graph of the Zeta function including it's critical line of nontrivial zeros. The Euler-Mascheroni constant is related to the Zeta-like limit below. The equation below <u>subtract</u> an log term from a exponential term but other expressions can <u>add</u> them and they'll share the similar Euler-Mascheroni limit. If it's graphed in the complex plane as $\Sigma(1/(n^s-\log_n(s))$, assuming $\log_1(s)=0$ then it's graph will look like Zeta in the entire C. plane.

$$\sum_{n=1}^{\infty} \frac{1}{n^x - \text{Log}_n(x)} \approx \text{limit value}$$

Assuming $\log_1(x)=0$ It's limit will be similar to the Euler-Mascheroni constant (0.577215...).

11. A new Alternating Euler Product for R(s)>0

People have formulated alternating Euler Products were it contains +,- series. However, I believe their not true Alternating Euler Products. To be a true alternating Euler Product we must also alternate between multiplication "x" and division "÷" along with +,-.

$$\prod{}^{(\times,\div)^p} \left(\frac{1}{1 - \frac{1}{p^s}} \right)^{(-1)^p}$$

Because multiplication or division requires two terms (i.e. a x b or a/b) we can't place a multiplication or division symbol before the first Prime term $(1/(1+(1/2^s)))$. After the first Prime term we associate division with Prime subtraction and for multiplication we use Prime addition. This +,- alternating series corresponds to an alternating multiplication "x" and division "÷" series. Conjecturally, this is a true alternating Euler Product!

$$\prod{}^{(\times,\div)^p} \left(\frac{1}{1-\frac{1}{p^s}} \right)^{(-1)^p} = \frac{1}{1+2^{-s}} \div \frac{1}{1-3^{-s}} \times \frac{1}{1+5^{-s}} \div \frac{1}{1-7^{-s}} \times \frac{1}{1+11^{-s}} \cdots$$

96

When R(s)>0 this new alternating Euler Product series is converging towards finite unique values similar to Zeta(2)= $\pi^2/6$.

$$\prod^{(x,\div)^p} \left(\frac{1}{1-\frac{1}{p^s}}\right)^{(-1)^p} = \frac{1}{1+2^{-s}} \div \frac{1}{1-3^{-s}} \times \frac{1}{1+5^{-s}} \div \frac{1}{1-7^{-s}} \times \frac{1}{1+11^{-s}} \cdots$$

Each alternating term in the solution is smaller than its predecessor. As stated before because multiplication or division requires two terms (i.e. a x b or a/b) we can't place a multiplication or division symbol before the first Prime term $(1/(1+(1/2^{\wedge}s)))$. The first term has a interesting distinction in the alternating series because the number "2" is the only even Prime Number while the rest in the alternating series are odd number Primes. The take away from this new alternating Euler Product is that we can not only have + and - alternating series but we can also have multiplication "x" and division "÷" alternating series coupled with the +,- alternating series for a new Alternating Euler Product series! Different x, ÷ alternating series may exist.

Alternating E. product = Alt. Zeta function

Just as the Zeta function is extended to greater than 0 as a alternating series it may be possible to extend the Euler Product to greater than zero by an alternating series of multiplication and division terms.

$$\sum \frac{(-1)^{n-1}}{n^s} = \prod_p (\times \div) \frac{(-1)^n}{\left(1 - \frac{1}{p^s}\right)}$$

It is to be determined if the two expressions above equal. Below show their parallels.

$$\sum_{n=1}^{\infty} \frac{(-1)^{n+1}}{n^z} = 1 - \frac{1}{2^z} + \frac{1}{3^z} - \frac{1}{4^z} + \cdots$$

$$\prod^{(\times \div)^p} \left(\frac{1}{1-\frac{1}{p^s}}\right)^{(-1)^p} = \frac{1}{1+2^{-s}} \div \frac{1}{1-3^{-s}} \times \frac{1}{1+5^{-s}} \div \frac{1}{1-7^{-s}} \times \frac{1}{1+11^{-s}} \cdots$$

Non-trivial zeros from the A. Euler product

Just as the alternating Zeta function (Dirichlet ETA function) computes nontrivial zeros it may be possible that the alternating
Euler Product below

$$\prod{}^{(x,+)^p} \left(\frac{1}{1-\frac{1}{p^s}} \right)^{(-1)^p} = \frac{1}{1+2^{-s}} \div \frac{1}{1-3^{-s}} \times \frac{1}{1+5^{-s}} \div \frac{1}{1-7^{-s}} \times \frac{1}{1+11^{-s}} \cdots$$

computes the same nontrivial zeros of the
Zeta function within the critical strip.

12. Hexagonal Zeta function

If you add the Zeta function to the alternating reciprocal of odd numbers it's new combined Dirichlet series has "hexagonal numbers(1,6,15,28,45...) in it's denominators.

$$1+\frac{1}{2^3}+\frac{1}{3^3}+\frac{1}{4^3}+\frac{1}{5^3}+\frac{1}{6^3}+\frac{1}{7^3}+\frac{1}{8^3}\cdots \quad \zeta(3)$$

$$+\quad 1-\frac{1}{3^3}+\frac{1}{5^3}-\frac{1}{7^3}+\frac{1}{9^3}-\frac{1}{11^3}+\frac{1}{13^3}-\frac{1}{15^3}\cdots \quad \frac{\pi^3}{32}$$

$$\frac{2}{1}+\frac{19}{6^3}+\frac{152}{15^3}+\frac{279}{28^3}+\frac{854}{45^3}+\frac{1115}{66^3}+\frac{2540}{91^3}+\frac{2863}{120^3}\cdots$$

Multiply their denominators(ie. 2x3,3x5,4x7..)
<u>Hexagonal Dirichlet series</u>

$$\zeta(3)+\frac{\pi^3}{32}=2.171003\cdots=\frac{2}{1}+\frac{19}{6^3}+\frac{152}{15^3}+\frac{279}{28^3}+\frac{854}{45^3}+\frac{1115}{66^3}+\frac{2540}{91^3}+\frac{2863}{120^3}\cdots$$

Notice how the leading coefficients in the denominators are hexagonal numbers (1,6,15,28,45..)

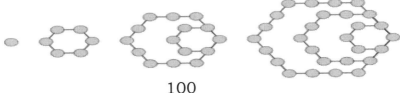

100

$$\text{Zeta}(1)+ \text{pi}/4$$

$$1 + \frac{1}{2} + \frac{1}{3} + \frac{1}{4} + \frac{1}{5} + \frac{1}{6} + \frac{1}{7} + \frac{1}{8} + \frac{1}{9} \cdots \quad \zeta(1)$$

$$+ \quad 1 - \frac{1}{3} + \frac{1}{5} - \frac{1}{7} + \frac{1}{9} - \frac{1}{11} + \frac{1}{13} - \frac{1}{15} + \frac{1}{17} \cdots \quad \frac{\pi}{4}$$

$$2 + \frac{1}{6} + \frac{8}{15} + \frac{3}{28} + \frac{14}{45} + \frac{5}{66} + \frac{20}{91} + \frac{7}{120} + \frac{26}{153} \cdots$$

Notice how the denominator in the new series generates hexagonal numbers (1,6,15,28,...). The numerators in the new combined divergent infinite series Zeta(1)+pi/4 are related to Diophantine relations! Below are first 9 numerators of the new infinite series. They share a unique Diophantine relationship.

Diophantine relations:

```
2 + 1 + 8 - 3 - 14 + 5 + 20 + 7 - 26 = 0
2 + 1 - 8 - 3 + 14 - 5 - 20 - 7 + 26 = 0
2 - 1 - 8 - 3 + 14 - 5 + 20 + 7 - 26 = 0
2 - 1 + 8 - 3 - 14 - 5 - 20 + 7 + 26 = 0
2 - 1 - 8 + 3 - 14 + 5 - 20 + 7 + 26 = 0
```

Wolfram generated a closed forms for the numerators.

$$a_n = (-1)^n \left(2(-1)^n n - n + (-1)^{n+1} \right) \text{ (for all terms given)}$$

101

The following are more combinations between the Zeta function and the reciprocal of
odd numbers and how they create hexagonal
numbers in the denominators.

$$\frac{\pi^2}{6} + \sum_{n=0}^{\infty} \frac{(-1)^n}{(2n+1)^2} = 2 + \frac{5}{6^2} + \frac{34}{15^2} + \frac{33}{28^2} + \frac{106}{45^2} \cdots$$

$$\zeta(3) + \frac{\pi^3}{32} = 2.171003\cdots = \frac{2}{1} + \frac{19}{6^3} + \frac{152}{15^3} + \frac{279}{28^3} + \frac{854}{45^3} + \frac{1115}{66^3} + \frac{2540}{91^3} + \frac{2863}{120^3} \cdots \approx \frac{3870028009\pi}{560056491}$$

$$\frac{\pi^4}{90} + \sum \frac{(-1)^n}{(2n+1)^4} = 2 + \frac{65}{6^4} + \frac{706}{15^4} + \frac{2145}{28^4} + \frac{7186}{45^4} \cdots$$

$$\zeta(5) + \frac{5\pi^5}{1536} = 2.0341100067\ldots = \frac{2}{1} + \frac{221}{6^5} + \frac{3368}{15^5} + \frac{15783}{28^5} + \frac{62174}{45^5} \cdots$$

$$\begin{array}{r} 1 + \dfrac{1}{2^5} + \dfrac{1}{3^5} + \dfrac{1}{4^5} + \dfrac{1}{5^5} \cdots \\[2mm] 1 - \dfrac{1}{3^5} + \dfrac{1}{5^5} - \dfrac{1}{7^5} + \dfrac{1}{9^5} \cdots \\[1mm] + \rule{6cm}{0.8pt} \\[1mm] \dfrac{2}{1} + \dfrac{221}{6^5} + \dfrac{3368}{15^5} + \dfrac{15783}{28^5} + \dfrac{62174}{45^5} \cdots \end{array}$$

102

Different variants

The Zeta function can also be expressed as a alternating Zeta function. The reciprocal of odd numbers can be an alternating sum or all addition sum. Their are different sign combinations but all will generate hexagonal numbers in their denominators. The following are random combinations of Hex. # Zetas.

$$\sum \frac{1}{n^2} + \sum \frac{1}{(2n-1)^2} = \frac{\pi^2}{6} + \frac{\pi^2}{8} = \frac{2}{1} + \frac{13}{6^2} + \frac{34}{15^2} + \frac{65}{28^2} + \frac{106}{45^2} \cdots = \sum_{n=1}^{\infty} \frac{n(5n-4)+1}{n^2(2n-1)^2} = \frac{7\pi^2}{24}$$

$$\sum_{n=1}^{\infty} \frac{(-1)^{n-1}}{(n^3)} + \sum_{n=0}^{\infty} \frac{(-1)^n}{(2n+1)^3} = 2 - \frac{35}{6^3} + \frac{152}{15^3} - \frac{407}{28^3} + \frac{854}{45^3} - \frac{1547}{66^3} \cdots$$

$$\sum_{n=1}^{\infty} \frac{(-1)^{n-1}}{(n^3)} + \sum_{n=0}^{\infty} \frac{(-1)^n}{(2n+1)^3} = 2 - \frac{35}{6^3} + \frac{152}{15^3} - \frac{407}{28^3} + \frac{854}{45^3} - \frac{1547}{66^3} \cdots$$

We can make different combinations of Hex-agonal Zeta functions. We can also study them in the complex plane. Below is one variant that can be explored in the complex plane

$$\varsigma(s) + \sum_{n=0}^{\infty} \frac{(-1)^n}{(2n+1)^s}$$

Zeta function + Dirichlet beta function = Hexagonal Zeta function

The Hexagonal Zeta function is a true L-function in the context it has a summed Euler Product & Summed functional equation.

$$\zeta(s) + \beta(k)$$

$$\sum \frac{1}{n^s} + \sum_{n=0}^{\infty} \frac{(-1)^n}{(2n+1)^s}$$

$$\prod_{p=prime} \frac{1}{(1-p^{-s})} + \prod_{p>2} \frac{1}{1-(-1)^{(p-1)/2}p^{-s}};$$

$$\zeta(s) = 2^s \pi^{s-1} \sin\left(\frac{s\pi}{2}\right) \Gamma(1-s)\zeta(1-s) + \beta(1-s) = \left(\frac{\pi}{2}\right)^{-s} \sin\left(\frac{\pi}{2}-s\right) \Gamma(s)\beta(s)$$

$$\zeta(s) + \beta(1-s) = 2^s \pi^{s-1} \sin\left(\frac{s\pi}{2}\right) \Gamma(1-s)\zeta(1-s) + \left(\frac{\pi}{2}\right)^{-s} \sin\left(\frac{\pi}{2}-s\right) \Gamma(s)\beta(s)$$

Applications of the hexagonal Zeta function

As a reminder, it's called a hexagonal Zeta function because the combined Zeta Dirichlet series and Dirichlet beta produces a combined hexagonal number Dirichlet series. The hexagonal numbers are in it's denominators. The following is one example.

$$+ \frac{\begin{matrix} 1+\dfrac{1}{2^3}+\dfrac{1}{3^3}+\dfrac{1}{4^3}+\dfrac{1}{5^3}+\dfrac{1}{6^3}+\dfrac{1}{7^3}+\dfrac{1}{8^3}\cdots & \zeta(3) \\[2mm] 1-\dfrac{1}{3^3}+\dfrac{1}{5^3}-\dfrac{1}{7^3}+\dfrac{1}{9^3}-\dfrac{1}{11^3}+\dfrac{1}{13^3}-\dfrac{1}{15^3}\cdots & \dfrac{\pi^3}{32} \end{matrix}}{\dfrac{2}{1}+\dfrac{19}{6^3}+\dfrac{152}{15^3}+\dfrac{279}{28^3}+\dfrac{854}{45^3}+\dfrac{1115}{66^3}+\dfrac{2540}{91^3}+\dfrac{2863}{120^3}\cdots}$$

Study it's nontrivial zeros

When the Hexagonal Zeta function is studied in the complex plane we can study the value of it's nontrivial zeros and see how they relate to the distribution of Prime Numbers. Because we have many different combinations of Zeta and Dirichlet beta series each represent a different L-function. As a result we have a new family of Hexagonal L-functions.

13. The Sawtooth Zeta function

The singularity at Zeta(1) is undefined. <u>What if we can construct a new Zeta function version were Zeta(1) is defined?</u> The following explores treating Zeta(1) as a Sawtooth wave pattern and building a new Zeta function around the Sawtooth wave pattern.

$$Sawtooth = \frac{2}{\pi}\sum_{n=1}^{\infty}\frac{1}{n}\sin(n\omega)$$

$$= \frac{2}{\pi}\left(\sin(\omega)+\frac{1}{2}\sin(2\omega)+\frac{1}{3}\sin(3\omega)+\frac{1}{4}\sin(4\omega)+\frac{1}{5}\sin(5\omega)...\right)$$

We can raise the numbers to higher powers to behave like the Zeta function >1 & that will equal a Sawtooth Product over the Primes.

$$\frac{2}{\pi}\sum_{n=1}^{\infty}\frac{1}{n^s}sin(nw) = \frac{2}{\pi}\prod_{p}(\frac{1}{1-p^{-s}})sin(pw)$$

106

The beauty of this new sawtooth interpretation of Zeta(1), the harmonic series we can associate a wave form with it. The Sawtooth wave is related to the harmonic series.

$$Sawtooth = \frac{2}{\pi} \sum_{n=1}^{\infty} \frac{1}{n} \sin(n\omega)$$

$$= \frac{2}{\pi}\left(\sin(\omega) + \frac{1}{2}\sin(2\omega) + \frac{1}{3}\sin(3\omega) + \frac{1}{4}\sin(4\omega) + \frac{1}{5}\sin(5\omega)... \right)$$

Graphing Positive integers of Zeta

For the first time we can associate a waveform with the positive even and odd values of Zeta.

Graphing equation for pi^2/6

$$\frac{2}{\pi} \sum_{n=1}^{\infty} \frac{1}{n^2} \sin(nw) = \frac{2}{\pi} \prod_{p} \left(\frac{1}{1 - p^{-2}} \right) \sin(pw)$$

Graphing equation for Zeta(3)

$$\frac{2}{\pi} \sum_{n=1}^{\infty} \frac{1}{n^3} \sin(nw) = \frac{2}{\pi} \prod_{p} \left(\frac{1}{1 - p^{-3}} \right) \sin(pw)$$

107

Besides the critical line of the Zeta function the values of the positive Even and odd numbers have mysteries.

We don't understand why "pi" shows up in the positive even integer values of Zeta.

Second, we don't know if the positive odd values of Zeta have closed forms.

By graphing the positive even or odd integers of Zeta as Fourier waves may shed new insight into their mysterious nature. What's also interesting is that the graphs will be based on Primes. The infinite sum waveform equation can also be expressed as a infinite sum wave Euler Product.

$$\frac{2}{\pi}\sum_{n=1}^{\infty}\frac{1}{n^s}sin(nw) = \frac{2}{\pi}\prod_{p}(\frac{1}{1-p^{-s}})sin(pw)$$

The Sawtooth-Zeta function greater than 1 can then be analytically continued to the rest of the complex plane like Riemann extended the original Euler Zeta function. The advantage of this Sawtooth-Zeta function & its analytic continuation is that <u>the harmonics series is defined as the Sawtooth wave pattern</u>.

A new Fourier wave Zeta function

The following list some of the highlights of the Fourier wave Zeta function

1. This new Sawtooth Zeta function equal or greater than 1 can be represented as Fourier waves. Graphing these waves may shed light into the behavior of the positive even and odd integer of Zeta

2. The analytic continuation of the Sawtooth Zeta function will describe a new critical strip(and nontrivial zeros).

3. Based on analytic continuation and reflection by the critical line this new Fourier wave expression of Sawtooth-Zeta can shed new light on the negative odd integers values of regular Zeta.

The Tetration Sawtooth Zeta-function

The Sawtooth Zeta-function can be analytically continued to the rest of the complex plane. Based on this new Fourier wave Zeta function, Mathematicians can also explore higher powers of Fourier Zeta similar to Zeta(s^s^s).

Expressing all L-functions as Fourier waves

Graphing the entire Zeta function (analytic continuation) as Fourier waves will show the beauty of the Riemann Zeta function in a new light. Based on this new Fourier wave Zeta type function we can expand it to other Dirichlet series L-functions.

What's interesting is that the Fourier wave expression of Zeta and L-function in generals brings them in the world of Fourier wave signal processing.

Comb filter Zeta function conjecture

The nontrivial zeros has harmonics to them. From a purely harmonics viewpoint the real and imaginary part of Zeta(1/2+yi) behave like a Comb filter. The real and imaginary part are like two identical waves in a time delay and that time delay forms the notches (nontrivial zeros). The sound of Zeta is Zeta(1/2+yi) as a Comb filter signal sound.

Comb filter   R(Zeta(1/2+yi))

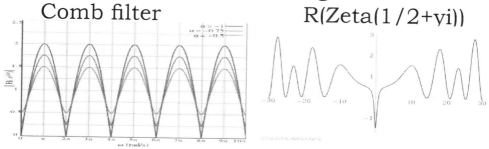

Relating Primes & signal processing

The Comb filter interpretation of the nontrivial zeros may shed light on primes. By definition a Prime number is a number that can only be divided by itself (i.e. 7/7 or 11/11) and 1. That is analogous to two identical signals in a time delay making Comb notches.

14. Zeta zeros as a Comb filter signal

In signal processing a comb filter occurs when two identical signals are in a time delay. The frequency response of a comb filter consists of a series of regularly spaced notches, giving the appearance of a comb. See the definition from Wikipedia.

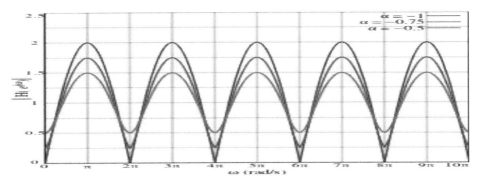

Where the comb filter dips down to zero are called the zeros of a comb filter.

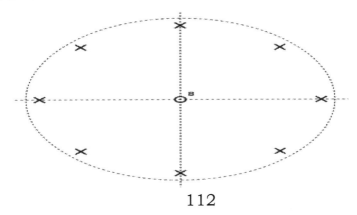

We can interpret the sound harmonic properties of the nontrivial zeros of Zeta at 1/2+yi as a comb filter. The real and imaginary part are analogous to two identical signals in a time delay like a Comb filter. The nontrivial zeros are analogous to the notches (zeros) of a comb filter.

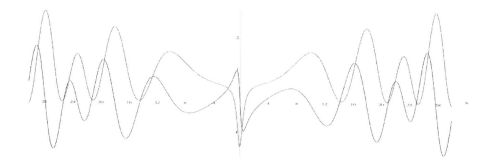

Apply the science of Comb filter signal processing to the critical line Zeta(1/2+yi). It's not just about the nontrivial zeros (14i,21i,..) but the waveform associated with the nontrivial zeros. If we want to hear the true sound of the nontrivial zeros then treat the waveform of Zeta(1/2+yi) as a Comb filter signal.

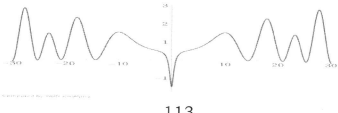

L-function Comb filters signals

The family of known L-functions describes different comb filters. For instance the real and imaginary waveforms of their critical lines represent two identical signals of a comb filter. The time delay (analogous to the real and imaginary wave) of their waveforms match the comb filter notices (nontrivial zeros).

The symmetric cube L-function(real part)

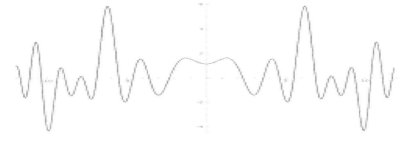

The Dedekind L-function(real part)

This sections shows how zeros of L-functions relate to comb filter signal processing. The following shows how Prime Numbers relate to signal processing.

Quantization(signals) and Prime Numbers

Wikipedia defines as "Quantization, in mathematics and digital signal processing, is the process of mapping input values from a large set (often a continuous set) to output values in a (countable) smaller set, often with a finite number of elements."

Graph of Quantization (signal processing)

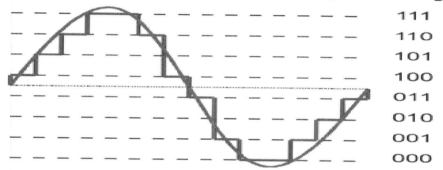

The smooth curve represent the ideal curve and the step curve represent the quantization.

Wikipedia describes Quantization error as the "The difference between an input value and its quantized value (such as round-off error) is referred to as quantization error."

The following is another graph of Quantization in signal processing.

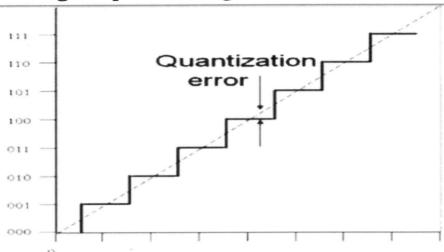

Notice the Signal processing Quantization graph has perfect correlation to the Prime count step function & Zeta zeros.

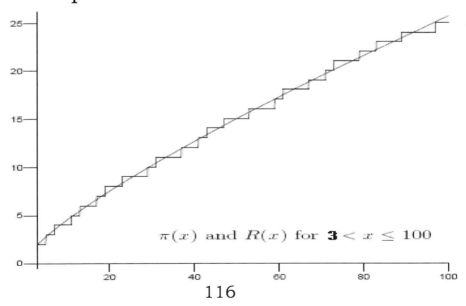

$\pi(x)$ and $R(x)$ for $3 < x \le 100$

The following is a comparison of the Zeta zeros-Prime step function and the Quantization signal error graph.

Perfect signal vs ideal

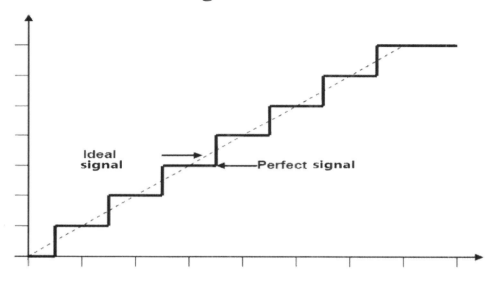

Prime count function vs Zeta zeros

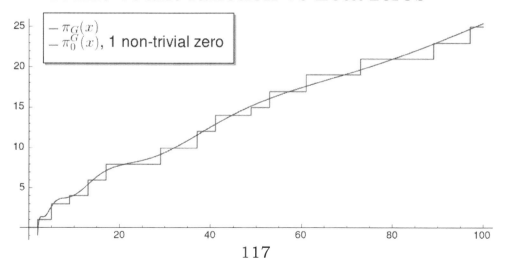

Primes & Zeta zeros signal relationship

Based on this research the distribution of Primes (step function) represent a perfect signal. The nontrivial zeros of Zeta related to Comb filter signals(zeros of comb filters) performs Quantization errors adjustment to the perfect Prime step function signal.

Connection to Random matrix theory

Because the nontrivial zeros of Zeta behave like energy levels inside heavy atomic nuclei it suggests that their are two fundamentally different energy levels. Meaning their is a perfect (ideal) eigenvalue energy levels like the Prime count step function. The energy levels we observe in heavy atomic nuclei are like Quantization error energy levels compared to the perfect (ideal) energy levels like Zeta zeros to the Prime step function.

Default Quantization tables and RMT

It was surprising how the nontrivial zeros relate to Random Matrix Theory. However, now that we see the signal processing quantization error of the nontrivial zeros we can see how they relate to Random Matrix theory.

Below is a image of a matrix that could be used in Random Matrix Theory(RMT).

$$\begin{pmatrix} 2 & -1 & 0 & 0 & -1 & 0 \\ -1 & 3 & -1 & 0 & -1 & 0 \\ 0 & -1 & 2 & -1 & 0 & 0 \\ 0 & 0 & -1 & 3 & -1 & -1 \\ -1 & -1 & 0 & -1 & 3 & 0 \\ 0 & 0 & 0 & -1 & 0 & 1 \end{pmatrix}$$

Below is a Default Quantization table used in signal processing(image processing).

This is an example of DCT coefficient matrix:

$$\begin{bmatrix} -415 & -33 & -58 & 35 & 58 & -51 & -15 & -12 \\ 5 & -34 & 49 & 18 & 27 & 1 & -5 & 3 \\ -46 & 14 & 80 & -35 & -50 & 19 & 7 & -18 \\ -53 & 21 & 34 & -20 & 2 & 34 & 36 & 12 \\ 9 & -2 & 9 & -5 & -32 & -15 & 45 & 37 \\ -8 & 15 & -16 & 7 & -8 & 11 & 4 & 7 \\ 19 & -28 & -2 & -26 & -2 & 7 & -44 & -21 \\ 18 & 25 & -12 & -44 & 35 & 48 & -37 & -3 \end{bmatrix}$$

119

Noise quantization of Primes & Zeros

The noise analogy of the Prime Numbers and nontrivial zeros can be capture in Noise quantization. What Riemann discovered with his Zeta function was a purely mathematical model of signal processing. The vast number of L-functions types based on the Zeta function are fundamentally based on signal processing quantization.

We can apply the difference between
True Prime and R(x)
to quantized signals of original & quantized.

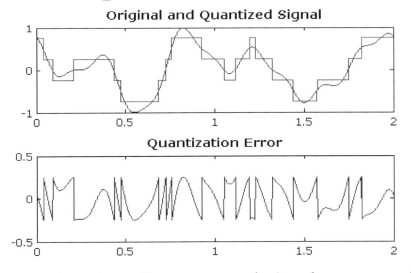

The Quantization Error graph is the same is the Error term between the true Prime value and the Prime count estimate.

L-function signal processing

The take away from this research is that Bernhard Riemann discovered the first pure mathematical model of signal processing. The way the nontrivial zeros behave compared to the true prime value and the difference between them is perfectly equivalent to signal processing were you have the original signal and quantized signal and the Quantization Error between them.

By treating the critical lines of L-functions as signals we can apply the science of signal processing (Noise quantization, Quantization errors, etc.) to study the distribution of Primes better and to possibility build better signal quantization technology for real world applications.

The future of L-functions is in the application of real world signal processing. It can be a new field called L-function signal processing.

15. <u>A interesting non-Zeta tetration function</u>

The following simple looking tetration equation has amazing Comb filter signal mathematical properties. Surprisingly it generates sophisticated 3D waveforms from the simple non-Zeta tetration equation.

$$1/(1-(1/x)^{\wedge}x^{\wedge}x) \qquad 1/(1-(1/s)^{\wedge}s^{\wedge}s)$$

3D wave forms   Represent 3D waveforms

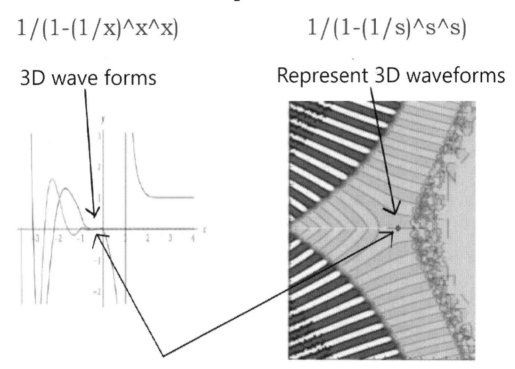

Arrows show where the 3D waveforms will appear in the 3D plot of $1/(1-(1/x)^{\wedge}x^{\wedge}x)$ over the reals & how it relates to its C. graph.

3D plot of $1/(1-(1/x)^{x^x})$

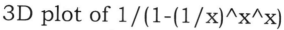

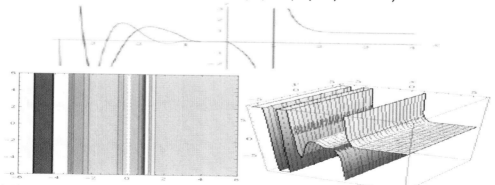

The 3D plot appears unremarkable until you plot it at different ranges and then amazing 3D waveforms appear!

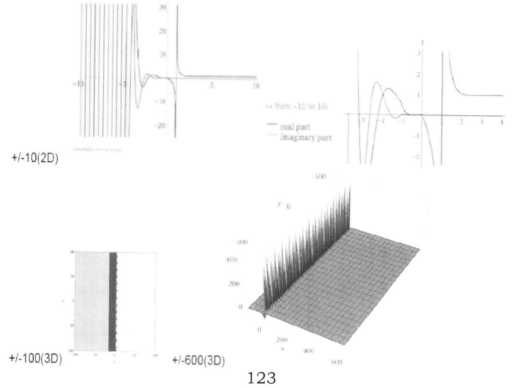

+/-10(2D)

+/-100(3D)   +/-600(3D)

Using Wolfram alpha at -600 to 600 for the 3D plot of 1/(1-(1/x)^x^x) we have the following 3D waveforms for the graph.

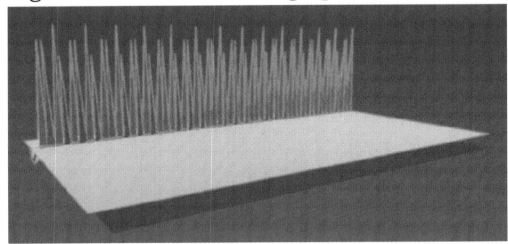

Graphs using Wolfram Mathematica

3DPlot 1/(1-(1/x)^x^x)) at the range of -950 to 950 using Wolfram mathematica

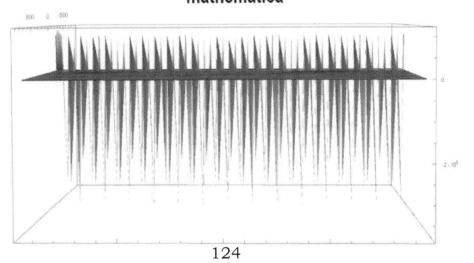

124

The 3D waveforms may appear like Random patterns but certain 3D plot reveals connections to Comb filter signal waveforms.

Plot3D[Re[(1 - (x^(-1))^x^x)^(-1)], { z, -600, 600}, {x, -600, 671}]

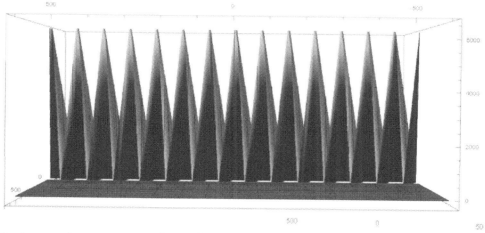

Below is a graph of Wikipedia of a Forwardfeed Comb filter signal.

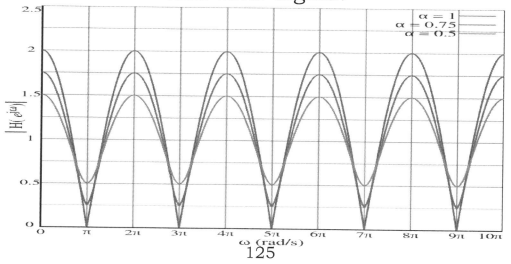

The following is another similarity between a 3D waveform generated by the new function 1/(1-(1/x)^x^x) and Comb filter signals.

Plot3D[Re[(1 - (x^(-1))^x^x)^(-1)], { z, -600, 600}, {x, -600, 1055}]

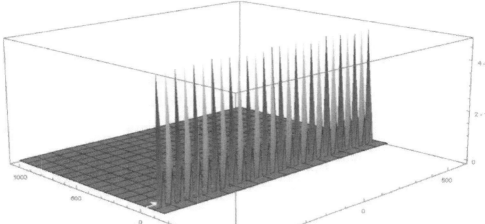

Below is a feedback Comb filter signal from Wikipedia. It's similar to the graph above.

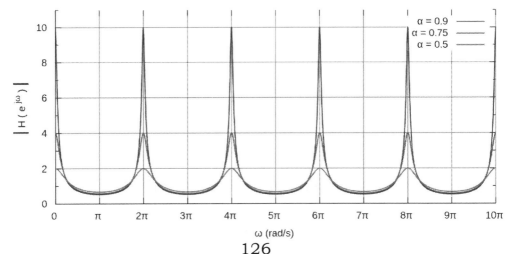

The following are more exotic waveforms the 3D plot of 1/(1-(1/x)^x^x) generates.

Plot3D[Re[(1 - (z^(-1))^z^z)^(-1)], {z, -879, 721}, {x, -891, 895}]

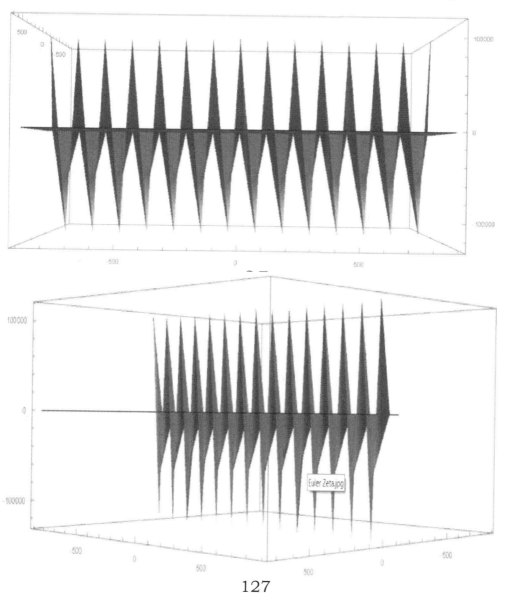

Plot3D[Re[(1 - (z^(-1))^z^z)^(-1)], {z, -900, 900}, {x, -920, 920}]

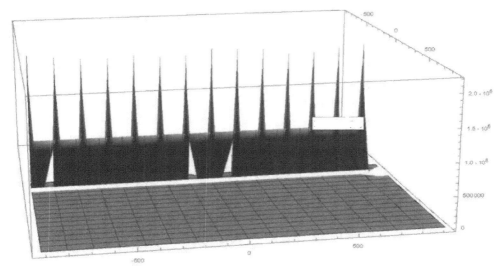

Plot3D[Re[(1 - (z^(-1))^z^z)^(-1)], {z, -911, 911}, {x, -923, 923}]

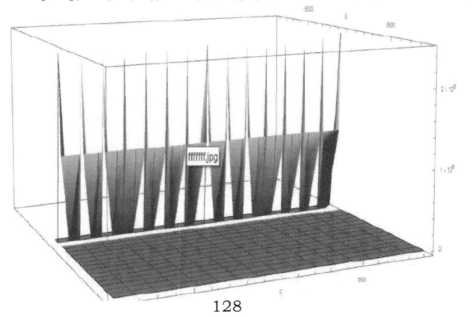

Plot3D[Re[(1 - (x^(-1))^x^x)^(-1)], { z, -600, 600}, {x, -600, 641}]

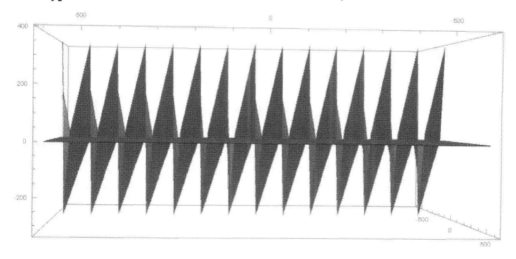

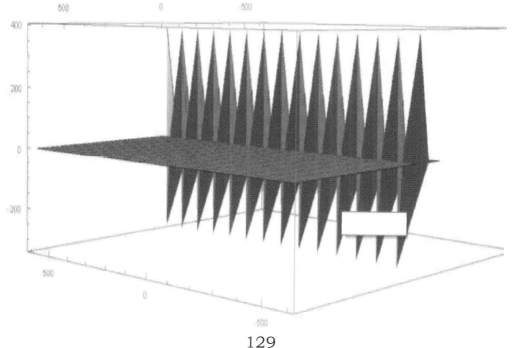

129

Plot3D[Re[(1 - (z^(-1))^z^z)^(-1)], {z, -811, 849}, {x, -615, 695}]

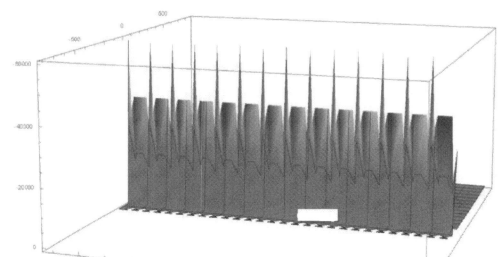

Plot3D[Re[(1 - (z^(-1))^z^z)^(-1)], {z, -847, 847}, {x, -861, 861}]

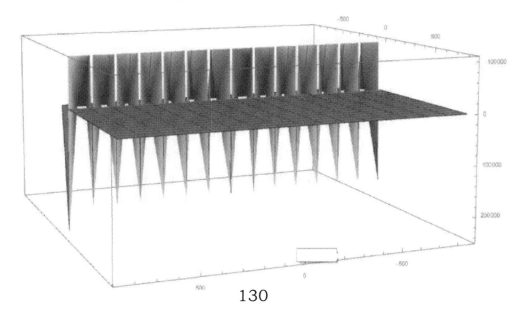

130

Plot3D[Re[(1 - (z^(-1))^z^z)^(-1)], {z, -800, 841}, {x, -781, 805}]

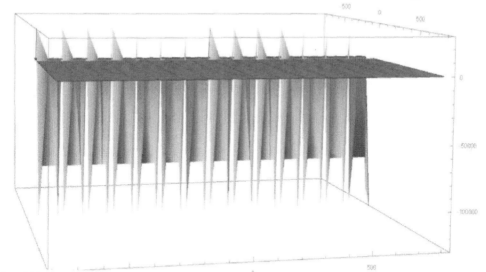

Plot3D[Re[(1 - (x^(-1))^x^x)^(-1)], { z, -600, 600}, {x, -600, 745}]

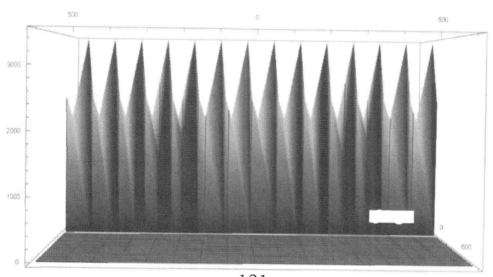

131

Plot3D[Re[(1 - (x^(-1))^x^x)^(-1)], { z, -600, 600}, {x, -600, 790}]

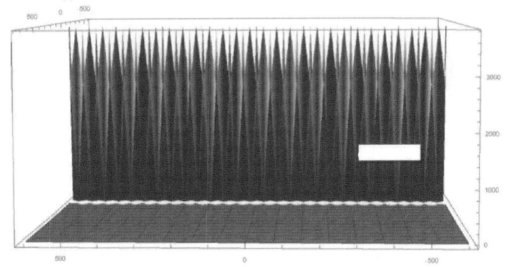

Plot3D[Re[(1 - (x^(-1))^x^x)^(-1)], { z, -600, 600}, {x, -600, 1090}]

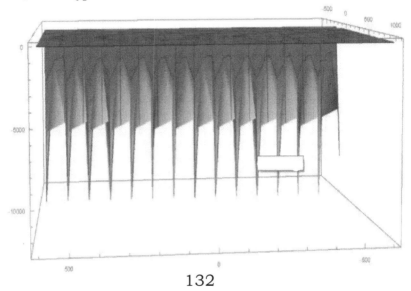

132

Plot3D[Re[(1 - (x^(-1))^x^x)^(-1)], { z, -600, 600}, {x, -600, 1065}]

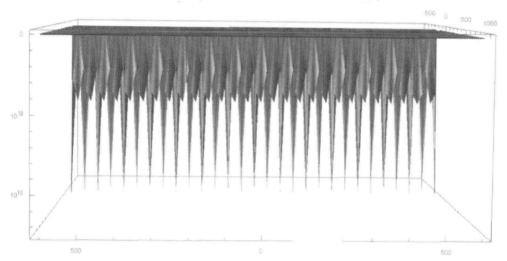

Plot3D[Re[(1 - (x^(-1))^x^x)^(-1)], { z, -600, 600}, {x, -600, 1064}]

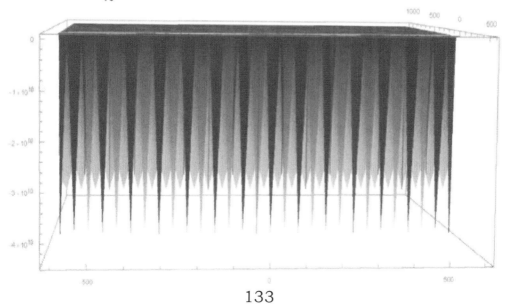

133

Extracting useful mathematics

Their is potentially an infinite number of different 3D waveform patterns the simple function 1/(1-(1/x)^x^x) can generate. You can take different plot ranges or combinations of plot ranges. Two examples from the graph show they have amazing similarities to Comb filter signal patterns. That implies that the 3D waveforms this new Tetration function generates is related to Comb filter signal patterns. The initial difficulty with the function 1/(1-(1/x)^x^x) was how to extract useful mathematics from it beside looking at the pretty 3D graph waveform pictures.

The solution on how to extract useful mathematical data from the new Tetration function was to study it's 3D waveforms & apply the following type of formulas. For example, the Forward feed filter & Feedback filter equations can be used as a template for study.

$$H\left(e^{j\Omega}\right) = 1 + \alpha e^{-j\Omega K}$$

$$H\left(e^{j\Omega}\right) = \frac{1}{1 - \alpha e^{-j\Omega K}}$$

Conjecturally, all the 3D waveforms generated by the new Tetration function $1/(1-(1/x)^{\wedge}x^{\wedge}x)$ can be represented in some form of the Forward feed or feedback filter equation. All the 3D Comb filter patterns from the equation are based on poles-zeros of a Comb filter signal.

The final take away of the new Tetration function is it's close similarity to L-functions. For example, notice it's 4 key features.

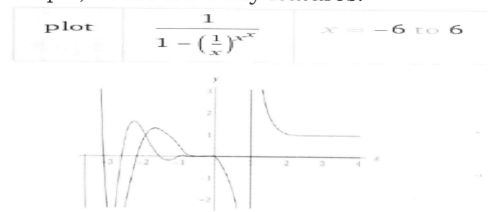

plot	$\dfrac{1}{1-\left(\frac{1}{x}\right)^{x^x}}$	$x = -6\ \text{to}\ 6$

1. Converges greater than 1.
2. It has a pole at 1
3. It has zeros(Comb filter poles-zeros) on a critical line(3D waveform Comb filter signal)
4. It has trivial zeros up to -15.

16. <u>Discovery of a new Elliptic curve form</u>

The Complex graph of 1/(1-(1/s)^s^s) show a E. curve pattern embedded in it's C. graph.

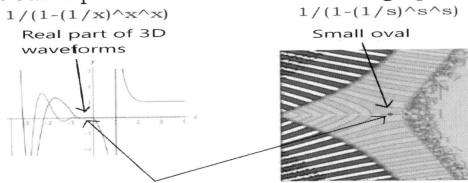

1/(1-(1/x)^x^x)

Real part of 3D waveforms

1/(1-(1/s)^s^s)

Small oval

Below is the graph of 1/(1-(1/s)^s^s). You'll notice it has Modular Lambda patterns in a arc and a small oval on the negative side like a Elliptic curve y^2=x^3-x.

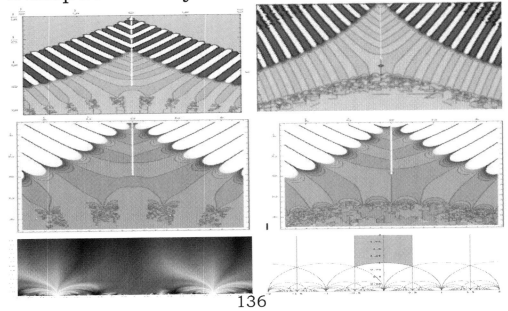

When studying the open arc on the complex graphs they are Modular Lambda function patterns. Notice how it forms an open arc and a small oval on the negative side similar to the graph of the Elliptic curve y^2=x^3-x.

y^2=x^3-x   1/(1-(1/s)^s^s)

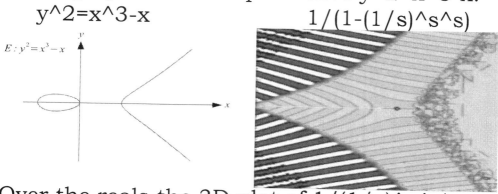

Over the reals the 3D plot of 1/(1/x)^x^x) has a <u>finite</u> repeating waveform patterning of Comb filter zeros. The arc of Modular lambda patterns for it's complex graph stretches out to <u>infinity</u> for infinite solutions.

<u>Coincidence or discovery?</u>

It's interesting how the pattern in the complex graph of 1/(1-(1/s)^s^s) has a <u>arc</u> of Modular lambda functions and a <u>small oval</u> on the negative side that corresponds to a finite repeating patterns of Comb filter zeros. If this Tetration Elliptic curve conjecture is proven to be true then it would be exciting!

Discovery of a new Elliptic curve form?

Let's say the Modular lambda functions and oval on the complex graph of 1/(1-(1/s)^s^s) makes a new type of Elliptic curve. What would that mean? Normally when we think of E. curves we think of a single curved line and a close loop oval. Could their exist an Elliptic curves based on a Modular Lambda function?

y^2=x^3-x   1/(1-(1/s)^s^s)

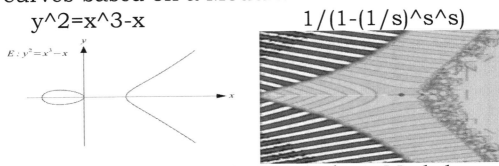

The graph of 1/(1-(1/s)^s^s) show Modular Lambda patterns. Modular Lambda functions are related to Elliptic curves. Is it a coincidence that this Tetration function that has λ(τ) functions has a Elliptic curve pattern in it's graph? Following is a crude new equation.

$$λ(τ)^2= λ(τ)^3-λ(τ)$$

We have two different Tau function analogous to y^2 and x^3-x and have them equal. They form a new type of Elliptic curve like in the graph of 1/(1-(1/s)^s^s)

138

Elliptic Tau curve conjecture

What if Elliptic curves using plain curve lines such as y^2=x^3-x is just the most basic form? What if the curved lines and ovals can be replaced by more sophisticated structures? For example the plot of y^2 can be more than a simple parabola but a sophisticated structure. The same for x^3-x. <u>What if the complex graph below(right) is a more sophisticated Elliptic curve?</u> Below is a crude concept of it's Elliptic Tau curve structure.

y^2=x^3-x   λ(τ)^2= λ(τ)^3-λ(τ)

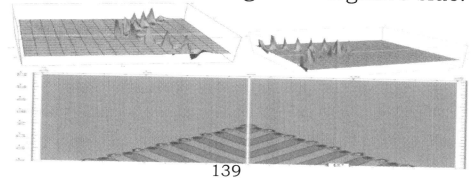

More graphs of 1/(1-(1/s)^s^s) show it has poles in a V array starting from negative side.

A tetration based Elliptic curve

The argument is that simple curved lines are just one form of a Elliptic curve. If the pattern on the complex graph of $1/(1-(1/s)^{s^s})$ is truly a Elliptic curve then it represents a more advance form of a Elliptic curve.

y^2=x^3-x   1/(1-(1/s)^s^s)

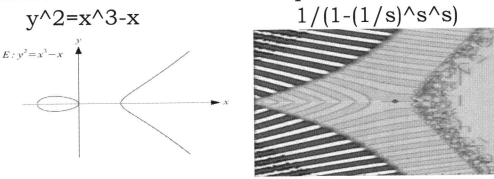

The similarity is uncanny. The arc in the tetration function consists of Modular Lambda functions in a Elliptic curve open arc with a infinite number of solutions. The oval on the negative side corresponds to the 3D wave forms that represent a finite repeating 3D waveform(Comb filter zeros or roots). <u>The goal is to determine if a new Elliptic curve equation can be found describes the conjectured Elliptic curve pattern found in the complex Tetration graph 1/(1-(1/s)^s^s).</u>

If it is truly a Elliptic curve then that will open up the door for a new realm of Elliptic curves, modular forms and L-functions related to Elliptic curves.

What's even more exciting is that just as a Torus corresponds to a Elliptic curve their would be a new type of Torus that corresponds to the Elliptic curve embedded within $1/(1-(1/s)^s{}^s)$. The goal is to extract that new sophisticated Elliptic curve as a stand alone equation if the Elliptic Tau curve conjecture is correct. This Elliptic curve embedded in $1/(1-(1/s)^s{}^s)$ is of Genus of 1 and is related to a Torus.

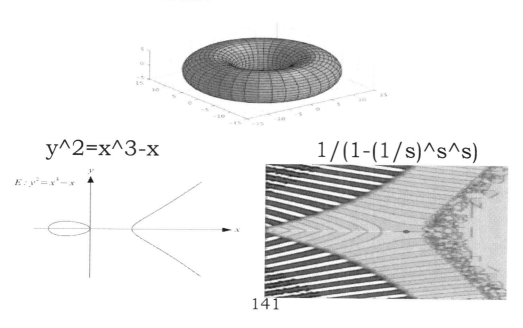

y^2=x^3-x

1/(1-(1/s)^s^s)

$E: y^2 = x^3 - x$

17. <u>Tetration L-functions for Hyper Elliptic curves</u>

If the tetration function 1/(1-(1/s)^s^s) makes a new type of Complex Elliptic curve embedded in a graph then what about a tetration Zeta function? Do we have a Elliptic curve in it? Let's graph Zeta as Zeta(s^s^s). The reason why we need 3 complex terms is that it correlates to the 3 complex terms in 1/(1-(1/s)^s^s). The graph of Zeta(s^s^s) behaves similar to the graph of (1-(1/s)^s^s).

plot	$\zeta\left((x + y\,i)^{(x+y\,i)^{x+y\,i}} \right)$	$x = -2$ to 2

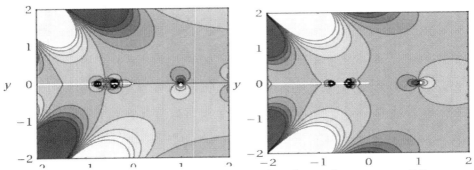

The Tetration Zeta(s^s^s) also have a V array of poles like (1-(1/s)^s^s).

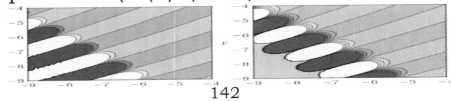

When we look at the complex graph of Zeta(s^s^s) it doesn't look exactly like a Elliptic curve of Genus 1 based on y^2=x^3-x. It resembles a Hyper elliptic curve C of genus 2. It has a open arc of M. Lambda patterns & the 2 circular parts and we can ignore Zeta(1).

Zeta(s^s^s)   C: y^2+h(x)y=f(x)

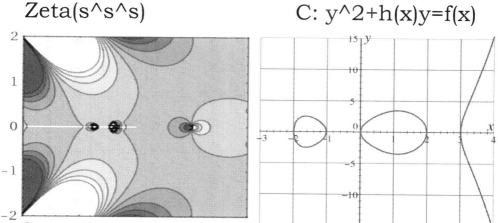

It's difficult to graph the Zeta function as Zeta(s^s^s) due to high computing requirements. It has an open arch of Modular Lambda function patterns on the right and 2 circular parts on the left like a hyper elliptic curve pattern were it looks like a Hyper Elliptic curve. If this conjecture is true then a stand alone Hyper elliptic curve equation can be extracted from the graph of Zeta(s^s^s) similar to the Elliptic curve form of y^2+h(x)y=f(x).

143

Tetration Zeta-function & Hyper Elliptic curve

Their are L-functions associated with Elliptic curves such as the Hasse–Weil zeta function. A Torus is related to a E. curve of Genus 1. If the complex graph of Zeta(s^s^s) has a Hyper Elliptic curve embedded in it then it will show that Tetration L-functions are related to Hyper Elliptic curves of Genus 2.

Purely Complex Elliptic curves!

Traditionally Elliptic curves have been studied using only real numbers(i.e. y^2=x^3-x). If this conjecture is true then we have purely complex Elliptic curves. Notice how the open arc of Modular lambda functions and the 2 circular structures looks like a Hyper Elliptic curve (ignoring Zeta(1).

Zeta(s^s^s)   y^2+h(x)y=f(x)

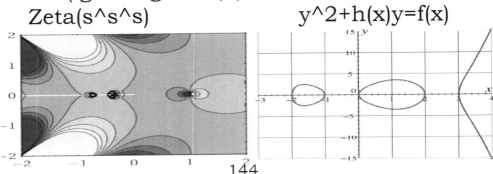

144

The interesting non-Zeta function has a Elliptic curve with a Genus of 1.

y^2=x^3-x   1/(1-(1/s)^s^s)

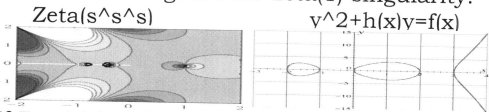

The arc on the graph on the right is a arc of
Modular Lambda patterns that stretch. The
oval on the negative side corresponds to the
finite repeating pattern (3D wave forms from
1/(1-(1/x)^x^x) from the previous chapters).
It has the properties of a E. curve. Regular L-functions are related to Elliptic curves
of Genus 1. When we graph Zeta(s^s^s) it
appears its graph has a Genus 2 Hyper E.
curve. We can ignore it's Zeta(1) singularity.

Zeta(s^s^s)   y^2+h(x)y=f(x)

If Zeta(s) is related to a E. curve Genus 1 and
Zeta(s^s^s) is related to a Hyper E. Curve Genus 2 then conjecturally, higher power towers
of ζ relates to higher Genus Elliptic curves.
We can visually check from its # of ovals.

145

18. A possible Quasi-crystal interpretation of the Riemann Zeta function

Image below is a picture of the Zeta function and the reciprocal of the Zeta function 1/Zeta(x+yi)

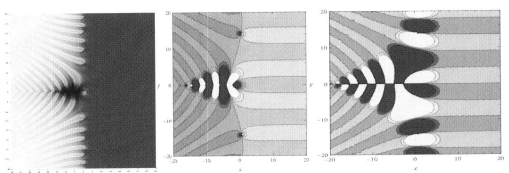

If we stretch our imagination notice how the graph of the Complex Riemann Zeta function looks amazingly similar to a diffraction pattern in physics!

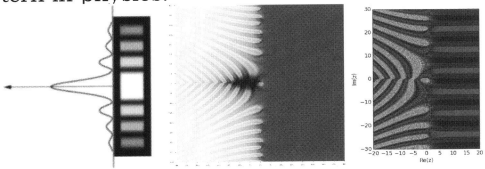

The horizontal patterns on the right of the
Zeta graph looks like the black and white interference pattern. The black mountain peak
at the center of the graph of the Zeta function
looks like a Airy Disk in 2D. The right side of
the Complex Zeta graph is acting like a light
source analogous to a light source in a Airy
disk diffraction pattern. The critical line is
like a wall with a hole in it. The Airy diffraction pattern represents a circular disk that
can represent a diffraction quasi crystal.

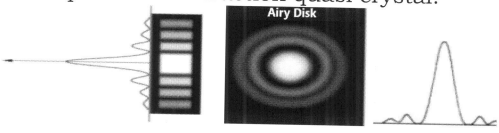

What if the 2D Zeta graph below is a type of
an Airy diffraction pattern? What if its only
showing a side image of a larger circular diffraction pattern for a Zeta Quasi-crystal?

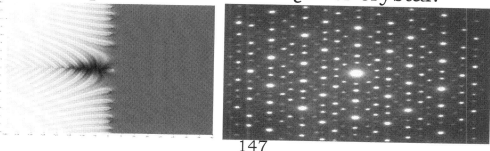

How do we determine the Quasicrystal structure of the nontrivial zeros?

It's possible the 2D graph of the Zeta function is showing a Airy disk.

We know that a Airy disk are circular but how can the critical line of Zeta be interpreted as a Circular?

Supercomputers are needed!

Given what we know about how Quasi crystal are made we can start with the nontrivial zero distribution.
Next, we can make the critical line and distribute it like the start of a Quasi-crystal. For example, look at the following quasi-crystal image

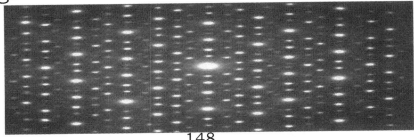

For the main discrete points patterns in lines in in the Quasi-crystal we can replace them with the nontrivial zero spacing (i.e. 14i, 21i, etc.). A supercomputer can try to determine Quasi-crystal tiling around the nontrivial zeros. Also it can introduce new point distributions within the patterns to help the quasi crystal work. By brute force it can try out different permutations of nontrivial zero point spacing in a circular Quasi crystal pattern to see what tiling can work. The final pattern will fit the mathematical description of a Quasi-crystal & be based entirely on Zeta zeros!

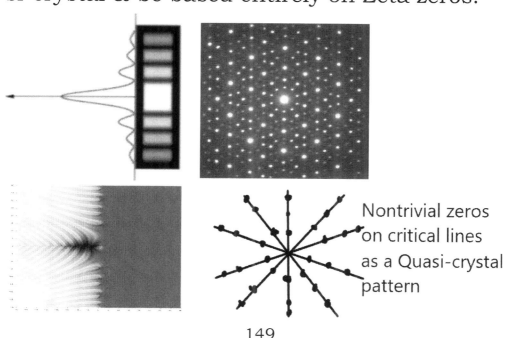

Nontrivial zeros on critical lines as a Quasi-crystal pattern

19. <u>Bonus Science and math discoveries</u>

<u>Mathematical secrets of the Neuron</u>

I've discovered a new mathematical pattern on the neuron pattern within the human brain. Mathematics is every where. This bonus section provides new insight into the mathematics of the human brain. More specifically it shows how neurons can be reduced to binary trees and then evolved into the sophisticated Neurons we see in the human brain. By understanding how the brain works we can build neural supercomputer that blend brute speed with pattern recognition and learning. The human brain is one of the most complicated systems. However, the complexity of the human brain can be reduce to fundamental behavior. We can build a basic human brain neural network based on binary $(0, 1)$ language and evolve it into the complex human brain. This mathematical model of the brain can be used either as a physical hardware system or a virtual processing were regular processors are used for this new neural software model.

The following is a breakdown of the neuron to its most basic binary state of 0,1. We can use it to build true neural computers for next generation computational mathematics. It shows how simple neurons can evolve into the complex neurons of the human brain. As current flows left or right that correlate to 0 or 1 binary digits. The typical switch of on or off is represented by left or right pathways along the neuron. Left and right corresponds to "0" & "1".

The mathematics of the neuron

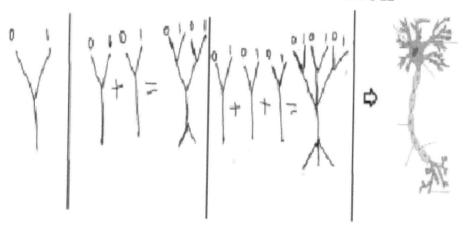

The complicated neuron is shown to be made from the summation of binary trees. It's possible to build a neural network based on the binary tree neuron model to form a brain.

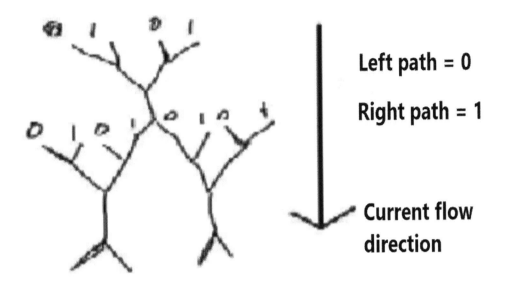

Left path = 0

Right path = 1

Current flow direction

The axon terminal of one binary neuron connects to the binary dendrite of another neuron that left and right path of the electrical signal represents a binary string. That sequence of left and right paths is recorded as a binary string of 1's and 0's in a memory trace. Using binary trees, we model binary structure neurons for neural computational mathematics. For simplicity we can still use a standard computer chip and use the neural architecture as a virtual neural processor to model the brain. This brain machine can compliment the Quantum computing era.

The mathematics of the human brain

The human brain is a powerful computer. In some ways even more advance that traditional supercomputer. Now imagine, taking one human brain and having it process in parallel with other human brains for a true neural supercomputer. The beauty is that the neural computer is not only fast but can it search for patterns and connections faster than any human does. Besides applications in Internet security, this neural computing can make for safer biz automation, Neuroscience biotech, neural consumer electronics applications like transforming the internet into the "Neuronet: A.I. search engine and human brain interface Internet".

Neural supremacy

The discovery of how the human brain works in binary can compliment Quantum computing. QC offers a brute force approach to computing but Neural computing based on my new brain discovery offers a brute force IQ approach for true superior artificial intelligence.

153

A new generalized Collatz conjecture

The following is a new pattern found when generalizing the Collatz conjecture. The patterns in this generalized form will make it easier for Mathematicians to derive a proof.

1. The even number leading column on the left (list below) generates only odd numbers.

2. The odd number leading column on the right (below) generates a sequence of even and odd numbers.

2n+1	1n+1
4n+1	3n+1
6n+1	5n+1
8n+1	7n+1
...	...

We can eliminate the "even leading number" column because it will only generate odd numbers. It can't reduce to the number one by even and odd steps.

The new Collatz family

Let's take the column of equations starting with odd numbers and include the "n/2" step to see if it leads to one. The pattern suggests all equations reduce to 1 but each equation has a different final loop number. All equations have a final loop in the form of "2n". A computer can test the "Generalized Collatz conjecture" up to certain values to see if loop patterns hold.

1n+1	**n/2 ---**	**Loops at 2**
3n+1	**n/2 ---**	**Loops at 4**
5n+1	**n/2 ---**	**Loops at 6**
7n+1	**n/2 ---**	**Loops at 8**
9n+1	**n/2 ---**	**Loops at 10**
...	**...**	**...**

The above pattern can be generalized into a "Generalized Collatz conjecture". Maybe mathematics is ready for the Generalized Collatz conjecture. It has patterns to help mathematicians derive a proof. Also the Collatz tree from each equation can be compared to other trees(via computer) to see if their are patterns between them. I call it the "Collatz forest"

Monstrous J invariant Space-CFT (JIS-CFT)

(Combining Monstrous Moonshine with ADS-CFT correspondence)

Their is a mysterious connection between the coefficient of the j-function

$$j(\tau) = \frac{1}{\bar{q}} + 744 + 196\,884\,\bar{q} + 21\,493\,760\,\bar{q}^2 + 864\,299\,970\,\bar{q}^3 + \ldots$$

and the Monster group in Group theory. It's an object that exists in 196883 dimension and how the following number of symmetries.

$$2^{46} \cdot 3^{20} \cdot 5^9 \cdot 7^6 \cdot 11^2 \cdot 13^3 \cdot 17 \cdot 19 \cdot 23 \cdot 29 \cdot 31 \cdot 41 \cdot 47 \cdot 59 \cdot 71$$

= 808,017,424,794,512,875,886,459,904,961
,710,757,005,754,368,000,000,000

$$\approx 8 \times 10^{53}.$$

Monstrous Moonshine

A mathematician proved a connection between the j-function and the Monster group via string. Other Mathematicians found new Moonshines related to Mock modular functions and K3 surfaces.

Connection between J-invariant and ADS

Imagine the J-invariant map like a 2D representations of hyperbolic 3D space. We can stack the J-invariants disks(maps) as slices in time like ADS maps it's time slices.

J-invariant   ADS-CFT

J Invariant Space-CFT (JIS-CFT)

We can construct a ADS-CFT type physics model but using the J-invariant map as space! The 2+1 surface of the J-invariants is like the surface on a Quantum field theory. The interaction on the 2D surface equate to interactions in the bulk. This new Monstrous J invariant space-CFT will relate to a unique string theory model like ADS-CFT relates to string theory. JIS-CFT may describe a new connection between the Monster group and the 2D J-invariant surface. Other Moonshines can be analogous JIS-CFT models.

New Elliptic curve research

If we take the following Elliptic curve
$$y^2=x^3-x$$
and evaluate the last term at higher odd powers in the form of

$$y^2 = x^3 - x^{2n-1}$$

something remarkable is unrevealed. The Elliptic curve flips. For the middle graph of $y^2=x^{\wedge}-x^3$ it's the same as $y^2=0$.

$y^2=x^3-x$ $\qquad\qquad$ $y^2=x^3-x^3$ $\qquad$ $y^2=x^2-x^{(5 \text{ or } 7 \text{ or } 9, \text{ or } 11...)}$

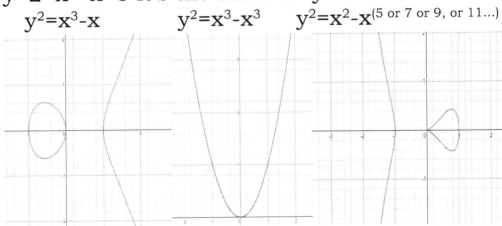

What's also interesting is that the oval in the reversed Elliptic curve takes on a tear drop shape and at higher odd powers it changes into a triangular shape.

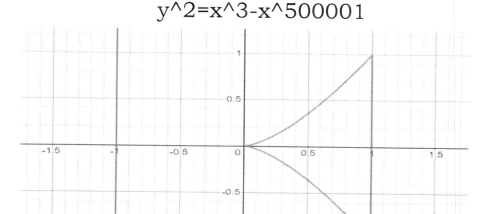

y^2=x^3-x^500001

Connection to the BSD conjecture

This new Elliptic curve research isn't proof
of BSD but shows how an Elliptic curve can
be evaluated at higher odd powers at one of
it's terms to study more of it's properties in
different ways. Those higher power E. curves
will have associated Modular forms, L-functions and Np/P graphs. It's provides more information on the behavior of the Elliptic curve
to help formulate a proof for BSD. Conjecturally, all E. curves can be evaluated to higher
odd powers at one of it's terms to reveal more
information on the Elliptic curves.

New Platonic solid discovery

Tetrahedron

Octahedron

Icosahedron

Cube

Dodecahedron

5,6,3,3,3,3,3,3,3,3,3,3,3,3...

The numbers above are regular polytopes that exists in the corresponding dimensions.

3,4,5,6,7,8,9,10,11,12,13.....

What if we took the number of regular polytopes in all dimensions and represented them as a decimal and fraction? For example,

0.05633333333...	= 169/3000
0.5633333333...	= 169/300
5.6333333333...	= 169/30
56.333333333...	= 169/3
563.33333333...	= 1690/3
5633.3333333...	=16900/3
56333.333333...	=169000/3
563333.33333...	= 1690000/3

Of course you can shift the decimal left or right but the number "169" and it's denominators have interesting geometric properties.

Properties of "169" and denominators

1. The numerator of the fraction representation of 5,6,3,3,3,3,3,3 is "169". The number 169 is a centered hexagonal number and center octagonal number.

2. The denominator <u>300</u> from 169/300= 0.563333... is a triangular number.

3. The denominator 30 from 169/30 = 5.633333... has special Platonic properties. The icosahedron and the dodecahedron are Platonic solids with 30 edges.

3. The denominator 3 from 169/3 = 56.33333... is a triangular number. Three of the five Platonic solids have triangular faces: the tetrahedron, the octahedron and the icosahedron. Also three of the 5 platonic solids have vertices were three faces meet: the tetrahedron, the hexahedron(cube) and the dodecahedron.

This section is just fun math observations with Platonics in decimal or fraction forms.

Mathematical discoveries when applying QM to General Relativity

Their has different ways to try to quantize General Relativity from treating the stress energy tensor as q-numbers or semi-classical gravity. Those approaches are dead ends.

I want to introduce a different way to apply QM to General Relativity. Let's treat the Einstein Field Equation as a complex conjugate wave function. At the heart of QM are complex numbers and Complex conjugate wave functions. For curiosity and exercise sake let's treat EFE itself as a Complex conjugate wave function. What's interesting is that if we multiply complex conjugate waves together we have new real values analogous to (a+bi) (a-bi)= a^2+b^2. Below is the EFE and we'll treat it like a Complex conj pair $\Psi^*\Psi$ wavefunction to retrieve new real values for GR.

$$R_{\mu\nu} - \frac{1}{2}g_{\mu\nu}R = \frac{8\pi G}{c^4}T_{\mu\nu}$$

We can multiply complex conjugate tensor & Scalar pairs to retrieve a new real value for linear EFE like $\Psi^*\Psi$. It will give a new exact EFE solution!

Discovery of Linear GR(LGR)

The <u>FOIL</u> method is applied to complex conjugate pair GR to retrieve real values based on (a+bi)(a-bi)= a^2+b^2 like linear Ψ*Ψ.

$$R_{(\mu\nu)^2+(\mu\nu)^2} - \frac{1}{2}g_{(\mu\nu)^2+(\mu\nu)^2}((R+Ri)(R-Ri)) = ((\frac{8\pi G}{c^4})^2 + (\frac{8\pi G}{c^4})^2)T_{(\mu\nu)^2+(\mu\nu)^2}$$

$$R_{(\mu\nu)^2+(\mu\nu)^2} - \frac{1}{2}g_{(\mu\nu)^2+(\mu\nu)^2}((R+Ri)(R-Ri)) = (\frac{128\pi^2(G^2)}{c^8})T_{(\mu\nu)^2+(\mu\nu)^2}$$

We have equal parts real & imaginary mass similar to (3+3i)(3-3i)=2(3^2).

$$R_{2(\mu\nu^2)} - \frac{1}{2}g_{2(\mu\nu^2)}(2R^2) = (\frac{128\pi^2(G^2)}{c^8})T_{2(\mu\nu^2)}$$

Notice the number 128 in the EFE constant!

What's so special about "128"?

What's so special about the number 128 in the new GR formula? What makes it so exciting is that the number "128" in the new EFE constant also appears in the energy coupling constant for the W-Boson (1/128). It's possible the number 128 in the new EFE could be describing a new Linear Graviton like the W-Boson! $\alpha \approx 1/128$

$$\text{at } Q^2 \approx m^2(W)$$

The new Linear General Relativity

Classical General Relativity is non-linear.

$$R_{\mu\nu} - \frac{1}{2}g_{\mu\nu}R = \frac{8\pi G}{c^4}T_{\mu\nu}$$

In order to make GR more compatible with QM the tensor and metrics was expanded to behave like linear wavefunction. For example you can multiply two wavefunctions together and retrieve a real solution based on based on (a+bi)(a-bi)= a^2+b^2)) like Ψ*Ψ.

The idea is that GR can be expanded to behave like (a+bi)(a-bi)=a^2+b^2. Also it's based on the assumption we have equal parts real and imaginary mass in our universe we can further condense the real solution. For example, (3+3i)(3-3i)=2(3^2). The following is a new GR equation with properties like a linear wavefunction Ψ*Ψ.

Below is a Linear GR formula like Ψ*Ψ.

$$R_{2(\mu\nu^2)} - \frac{1}{2}g_{2(\mu\nu^2)}(2R^2) = \left(\frac{128\pi^2(G^2)}{c^8}\right)T_{2(\mu\nu^2)}$$

Exact superposition solution

The question is can we obtain exact solutions to the new GR equation(i.e. $(3+3i)(3-3i)=2(3^2)$) that is structured like a complex conjugate pair wavefunction $\Psi^*\Psi$?

$$R_{2(\mu\nu^2)} - \frac{1}{2}g_{2(\mu\nu^2)}(2R^2) = (\frac{128\pi^2(G^2)}{c^8})T_{2(\mu\nu^2)}$$

<u>The answer may be YES!</u> For the new linear GR formulation above that is structured like $\Psi^*\Psi$ the sum of an exact and an approximate solution could equal a new exact solution.

2*(classic GR exact solution^2)

I haven't solved the equation above but if we look at the new GR equation above, everything is virtually the same as classical GR except we have the number "2" and the tensors and scalar are raised to the 2nd power. We could <u>take know exact solutions</u> from classical GR and multiply it by 2 and raise the tensor or scalar to the 2nd power and presumably <u>derive a new exact solution</u>.

Expanding GR to make it linear

There is nothing wrong with the EFE. The conjecture is that within the realm of QM the stress-energy Tensor isn't capture all possible forms of energy. For example, in QM we have complex numbers and Complex wavefunctions. Complex conjugate wavefunctions are multiplied together $\Psi^*\Psi$ to give new real values. It's based on $(a+bi)(a-bi)=a^2+b^2$. It has 4 parts: a,a,-bi,+bi. Therefore, if the stress-energy tensor of EFE is to be compatible with QM we need 4 different Stress-energy tensors based on $(a+bi)(a-bi)$ analogous to a complex conjugate pair wavefunction $\Psi^*\Psi$. Because EFE has to be real and need real values we can use the FOIL method. The new GR formulation needs 4 parts to the stress-energy tensors and it's Geometric side. It's assume that we have equal parts real mass and imaginary mass (i.e. $(3+3i)(3-3i)=2(3^2)$ we can express the new EFE as

$$R_{2(\mu\nu^2)} - \frac{1}{2}g_{2(\mu\nu^2)}(2R^2) = \left(\frac{128\pi^2(G^2)}{c^8}\right)T_{2(\mu\nu^2)}$$

LINEAR EFE

For QM wavefunctions we multiply complex conjugate wavefunctions $\Psi^*\Psi$ to retrieve real values. Likewise, let's express EFE as complex conjugate pairs & then multiply them to retrieve a new real value. The new value represent LINEAR (superimposed) distribution of the space-time geometry and the stress energy tensor like a smeared out wavefunction. All the matrices and scalars are the same values for their individual real parts (i.e. (3+3i)(3-3i)=3^2+3^2=2(3^2).

$$R_{(\mu\nu)^2+(\mu\nu)^2} - \frac{1}{2}g_{(\mu\nu)^2+(\mu\nu)^2}\big((R+Ri)(R-Ri)\big) = \left(\left(\frac{8\pi G}{c^4}\right)^2 + \left(\frac{8\pi G}{c^4}\right)^2\right)T_{(\mu\nu)^2+(\mu\nu)^2}$$

$$R_{(\mu\nu)^2+(\mu\nu)^2} - \frac{1}{2}g_{(\mu\nu)^2+(\mu\nu)^2}\big((R+Ri)(R-Ri)\big) = \left(\left(\frac{64\pi^2 G^2}{c^8}\right) + \left(\frac{64\pi^2 G^2}{c^8}\right)\right)T_{(\mu\nu)^2+(\mu\nu)^2}$$

$$R_{(\mu\nu)^2+(\mu\nu)^2} - \frac{1}{2}g_{(\mu\nu)^2+(\mu\nu)^2}\big((R+Ri)(R-Ri)\big) = \left(\frac{128\pi^2(G^2)}{c^8}\right)T_{(\mu\nu)^2+(\mu\nu)^2}$$

$$G_{2(\mu\nu^2)} = \left(\frac{128\pi^2(G^2)}{c^8}\right)T_{2(\mu\nu^2)}$$

Complex Special Relativity

To make GR on a fundamental level compatible with QM we need a big bang based on complex mass universe and it's complex conjugate pair mass universe oppose to a matter-antimatter universe. We need an analog of complex number for Relativity. The Special theory of Relativity has the ingredients to make up a complex number analog for Relativity that parallels the QM complex numbers:

Matter + Tachyons ---> (a+bi) in QM

Tachyons aren't forbidden. They can't slow down to the speed of light. If we add tachyons to real mass they make up Complex mass. We can have a complex conjugate pair.

Matter - Tachyons---> (a-bi) in QM

Complex General Relativity

We can extend Complex Special Relativity to complex GR. We can retrieve real values based on (a+bi)(a-bi) just like $\Psi^*\Psi$. (i.e. (3+3i) (3-3i)=2(3^2) were the EFE is expressed as

$$R_{2(\mu\nu^2)} - \frac{1}{2}g_{2(\mu\nu^2)}(2R^2) = \left(\frac{128\pi^2(G^2)}{c^8}\right)T_{2(\mu\nu^2)}$$

Complex conjugate pair big bang

Classical GR is conjecture to be incomplete in that it doesn't capture all possible energy forms. We may have had a complex conjugate pair big bang with equal parts real + imag & its complex conjugate pair. Those 4 over-all energy sections are similar to the 4 parts of a complex conjugate pair $\Psi^*\Psi$ or $(3+3i)(3-3i)=2(3^\wedge 2)$. Complex conjugate pair $E=2c^4m^2$ $=(mc^2+(mc^2)i)(mc^2-(mc^2)i)$ is used on complex GR. The reason why Complex GR is expressed like $(3+3i)(3-3i)=2(3^\wedge 2)$ is that we had a complex conjugate pair big bang of equal parts real matter & imag. matter.

$$R_{2(\mu\nu^2)} - \frac{1}{2}g_{2(\mu\nu^2)}(2R^2) = \left(\frac{128\pi^2(G^2)}{c^8}\right)T_{2(\mu\nu^2)}$$

COMPLEX MASS UNIVERSE(S)

The WGR formulation is based on a complex conjugate pair big bang. Conjecturally each universe may be quantum entangled. Meaning if a Graviton is emitted from real mass it's simultaneously emitted from it's imag. mass part(a+bi). Also Gravitons are emitted form it's complex conjugate pair(a-bi). When Gravitons emit they do so as Linearized Gravitons.

169

A new Total energy system

The following is the total energy and rest energy–momentum relation.

$$E^2 = p^2 c^2 + m_0^2 c^4,$$

What's interesting is that if we take the complex conjugate pair multiplication of energy+-momentum based on the FOIL method (a+bi) (a-bi) we have a formula similar to original Energy-momentum relation.

$$2(E^2) = 2(p^2 c^2) + 2(m^2 c^4)$$

$2E^2 \quad = (E+Ei)(E-Ei)$
$2(pc)^2 = (pc+pc(i))(pc-pc(i))$
$2m^2 c^4 = (mc^2+(mc^2)i)(mc^2-(mc^2)i)$

The new formulation represents the total energy in a complex conjugate pair system. It's based on the idea that our universe started from a complex conjugate pair big bang and not simply matter-antimatter big bang.

$$E=2m^2 c^4 = (mc^2+(mc^2)i)(mc^2-(mc^2)i)$$

Correspondingly, we can extend the new total energy system to the new linear GR formula. We apply the FOIL method (a+bi)(a-bi) or i.e. (3+3i)(3-3i)=2(3^2) to derive real values analogous to complex conjugate pair wavefunction multiplication Ψ*Ψ.

$$G_{2(\mu\nu^2)} = \left(\frac{128\pi^2(G^2)}{c^8}\right)T_{2(\mu\nu^2)}$$

$$R_{2(\mu\nu^2)} - \frac{1}{2}g_{2(\mu\nu^2)}(2R^2) = \left(\frac{128\pi^2(G^2)}{c^8}\right)T_{2(\mu\nu^2)}$$

The FOIL method to Linearize GR

To reconcile GR & QM their mathematics 1st and foremost must be compatible. Meaning GR has to have a linearized form, use complex numbers just like QM. I believe it's impossible to reconcile GR & QM without having them use similar mathematics. The only solution to make GR superposition(linear) & complex like QM is by the FOIL method (a+bi)(a-bi), i.e. (3+3i)(3-3i)=2(3^2).

171

An analogous wavefunction GR model

The premises of this new GR formulation is that it is treated like $\Psi^*\Psi$. Meaning it's properties are analogous to a wavefunction. I want to point that out because I'm not using the actual language of QM $\Psi^*\Psi$ but something similar.(i.e. (3+3i)(3-3i)=3^2+3^2=2(3^2)).

$$G_{2(\mu\nu^2)} = (\frac{128\pi^2(G^2)}{c^8})T_{2(\mu\nu^2)} \approx \psi^*\psi$$

$$G_{2(\mu\nu^2)} = (\frac{128\pi^2(G^2)}{c^8})T_{2(\mu\nu^2)}$$

$$R_{2(\mu\nu^2)} - \frac{1}{2}g_{2(\mu\nu^2)}(2R^2) = (\frac{128\pi^2(G^2)}{c^8})T_{2(\mu\nu^2)}$$

A wavefunction version of GR

The new stress-energy tensor represent the components of a wavefunction. In which the stress-energy tensor is expressed as (a+bi)(a+bi) analogous two wavefunctions $\Psi^*\Psi$. The beauty of this approach is that it's not modifying GR but expanding it. We're not using literal wavefunction QM language in GR but an analogous correlation with wavefunctions. Hence, it's called Wavefunction GR(WGR).

The Stress-energy tensor must have 4 parts to make it analogous to QM wavefunctions Ψ*Ψ. GR as (a+bi)(a-bi) is analogous to Ψ*Ψ.

$$\left(\frac{8\pi G}{c^4} + \frac{8\pi G}{c^4}i\right)\left(\frac{8\pi G}{c^4} - \frac{8\pi G}{c^4}i\right) = \frac{128\pi^2(G^2)}{c^8}$$

The new EFE are not literal wavefunctions but have properties analogous to wavefunctions. Not only do we see "128" in the new EFE but we see "128" in the new complex conjugate pair Einstein H. action: (a+bi)(a-bi).

$$S = \frac{\pi^2 G^2}{128} \int d^4x \sqrt{-(2(g^2))}2R^2 + 2(S_m[\psi, g_{\mu\nu}])^2$$

Numerical correlation!

What makes the new Quantum mechanical formulation of GR and the Einstein-Hilbert action so interesting is that we see the number "128". We see the same #128 in the W-Boson 1/128. Conjecturally the number "128" in the new GR/E-H is describing a Linear Graviton similar to the W-Boson!

$$\alpha \approx 1/128$$

$$at \ Q^2 \approx m^2(W)$$

The magic number "128"

Besides trying to solve the new EFE, what makes it so interesting is the number 128 that repeatably appears in various GR formulas. Their based on (3+3i)(3-3i)=2(3^2).

$$G_{2(\mu\nu^2)} = \left(\frac{128\pi^2(G^2)}{c^8}\right)T_{2(\mu\nu^2)}$$

$$S = \frac{\pi^2 G^2}{128}\int d^4x\sqrt{-(2(g^2))2R^2 + 2(S_m[\psi, g_{\mu\nu}])^2}$$

$$\int d^4x\sqrt{-((g+gi)(g-gi))}[L_{QFT} + \frac{\pi^2}{128}((R+Ri)(R-Ri))]$$

$$\alpha = (1/16\pi + 1/16\pi i)(1/16\pi - 1/16\pi i) = \frac{\pi^2}{128}$$

The reason why 128 is so interesting is that we see the number 128 in the W-Boson. NOTICE HOW THE ALPHA TERM π²/128 IS SIMILAR TO THE W-BOSON ALPHA 1/128.

$$\alpha = \frac{\pi^2}{128}$$

$$\alpha \approx 1/128$$
$$\text{at } Q^2 \approx m^2(W)$$

174

From Nonlinear GR to LINEAR GR

Nonlinear GR

$$R_{\mu\nu} - \frac{1}{2}g_{\mu\nu}R = \frac{8\pi G}{c^4}T_{\mu\nu}$$

describes a massless Spin 2 Graviton associated with it's stress-energy tensor. Their are problems trying to Quantize GR from it's classical form.

What I proposed is the stress-energy tensor in classical GR is incomplete in the context that it's not capturing all forms of energy in the language of QM. In particular Complex wavefunctions: $\Psi^*\Psi$. In order for GR to be compatible with QM, we must expand the Stress-energy tensor to be in the form of (a+bi)(a-bi) analogous to a complex conjugate pair QM : $\Psi^*\Psi$. In doing so GR becomes real again but in a new superimposed form analogous to a superposition QM wavefunction. This new <u>Linear GR</u>

$$R_{2(\mu\nu^2)} - \frac{1}{2}g_{2(\mu\nu^2)}(2R^2) = (\frac{128\pi^2(G^2)}{c^8})T_{2(\mu\nu^2)}$$

describes the Graviton with different properties that mediates the force of Gravity.

Superposition Linear Graviton interactions

Nonlinear GR is associated with a Spin 2 massless point particle. It would have point particle interaction.

$$R_{\mu\nu} - \frac{1}{2}g_{\mu\nu}R = \frac{8\pi G}{c^4}T_{\mu\nu}$$

Linear GR is associated with superimposed massless Spin 2 Gravitons(G) based on (G+Gi)(G-Gi)=2(G^2). Rather than having point interaction we would have a superimposed Gravitational interaction. The interaction won't be point-like but superposition like based on linear GR.

$$R_{2(\mu\nu^2)} - \frac{1}{2}g_{2(\mu\nu^2)}(2R^2) = \left(\frac{128\pi^2(G^2)}{c^8}\right)T_{2(\mu\nu^2)}$$

The coupling constant for the Graviton was retrieve from the complex path integral.

$$\int d^4x\sqrt{-((g+gi)(g-gi))}\left[L_{QFT} + \frac{\pi^2}{128}((R+Ri)(R-Ri))\right]$$

$$\alpha = (1/16\pi + 1/16\pi i)(1/16\pi - 1/16\pi i) = \frac{\pi^2}{128}$$

Alpha term π^2/128 is similar to the W Boson coupling constant(1/128). π^2/128 is describing a Linearized superposition Graviton.

176

Linear Gravitons scatter like the W Bosons

The Gravity alpha term π^2/128 isn't describing a massive Graviton but a superposition of Spin 2 massless Gravitons. That energy equals a mass particle like the W Boson.

Just as the Higgs helped with W Boson scattering at high energies the Higgs can also help with Linear Graviton scattering.

<u>Linear Graviton couping</u>   <u>W Boson</u>

$$\alpha = \frac{\pi^2}{128} \qquad \alpha \approx 1/128$$

$$\text{at } Q^2 \approx m^2(W)$$

Massless Superposition Gravitons

(Superposition energy state)

Linear GR is based on complex conjugate pair multiplication (i.e. (3+3i)(3-3i)=2(3^2)).

$$R_{2(\mu\nu^2)} - \frac{1}{2}g_{2(\mu\nu^2)}(2R^2) = \left(\frac{128\pi^2(G^2)}{c^8}\right)T_{2(\mu\nu^2)}$$

Imagine the Massless Spin 2 <u>G</u>ravitons in superposition (G+Gi)(G-Gi)=2(G^2). The coupling constant of π^2/128 is not referring to Graviton mass but it's superposition state.

177

We need both nonlinear GR & LGR

Just as Classical wave equations can describe certain large scale things and Schrodinger wavefunction can describe QM physics we also need to two types of GR.

1. We need the classical non-linear version of GR for Gravity waves (for example).

2. We also need a linear version of GR to describe Quantum Gravity effects.

It's not about replacing non-linear GR with a linear GR but suggesting we need both. When Quantum effects are not involved then we need non-linear GR but when Quantum effects are involved then we need linear GR.

Non-linear GR

$$R_{\mu\nu} - \frac{1}{2}g_{\mu\nu}R = \frac{8\pi G}{c^4}T_{\mu\nu}$$

Linear GR analogous to $\Psi^*\Psi$.

$$R_{2(\mu\nu^2)} - \frac{1}{2}g_{2(\mu\nu^2)}(2R^2) = \left(\frac{128\pi^2(G^2)}{c^8}\right)T_{2(\mu\nu^2)}$$

A new linearized Graviton

The core issues with the massless Spin 2 Graviton associated with classical GR is that classical GR is nonlinear. Fundamentally, the massless Spin 2 Graviton can't be the true particle of Gravity. If Gravitons exists then they'll have to be associated with a linearized version of GR. This book describes converting nonlinear GR into linear GR by expanding the tensors and scalar to behave like complex conjugate pair wavefunctions (a+bi) (a+bi) analogous two wavefunctions $\Psi*\Psi$. It is assumed we have equal parts real and imaginary mass. Because of that we can compact the tensor and scalar similar to (3+3i)(3-3i)=2(3^2).

$$R_{2(\mu\nu^2)} - \frac{1}{2}g_{2(\mu\nu^2)}(2R^2) = (\frac{128\pi^2(G^2)}{c^8})T_{2(\mu\nu^2)}$$

Notice how it retains the classical GR form except were multiplying it by "2" and raising the tensors to the 2nd power. What that mean we can have exact solutions for L. GR.

2*(classic GR exact solution^2)

The true version of the Graviton is based on the Linearized GR formulation above.

Solving the Complex EFE

$$G_{2(\mu\nu^2)} = \left(\frac{128\pi^2(G^2)}{c^8}\right)T_{2(\mu\nu^2)}$$

$$R_{2(\mu\nu^2)} - \frac{1}{2}g_{2(\mu\nu^2)}(2R^2) = \left(\frac{128\pi^2(G^2)}{c^8}\right)T_{2(\mu\nu^2)}$$

All matrices & scalars each have their same
real values similar to (3+3i)(3-3i)=3²+3²= 2(3²)

Squaring a diagonal matrix

Below is an example of a diagonal matrix ^7.

$$A^7 = \begin{bmatrix} 2^7 & 0 & 0 & 0 & 0 \\ 0 & 3^7 & 0 & 0 & 0 \\ 0 & 0 & 4^7 & 0 & 0 \\ 0 & 0 & 0 & 2^7 & 0 \\ 0 & 0 & 0 & 0 & 2^7 \end{bmatrix} = \begin{bmatrix} 128 & 0 & 0 & 0 & 0 \\ 0 & 2187 & 0 & 0 & 0 \\ 0 & 0 & 16384 & 0 & 0 \\ 0 & 0 & 0 & 128 & 0 \\ 0 & 0 & 0 & 0 & 128 \end{bmatrix}$$

Squaring the diagonal tensors will be straight
forward. We include the new Scalar(2) and
raise the tensors to a 2nd power & solve.

$$2\left(\begin{bmatrix} -\left(1-\frac{2GM}{rc^2}\right) & 0 & 0 & 0 \\ 0 & \left(1-\frac{2GM}{rc^2}\right)^{-1} & 0 & 0 \\ 0 & 0 & r^2 & 0 \\ 0 & 0 & 0 & r^2\sin^2\theta \end{bmatrix}\right)^2 \quad 2\left(\begin{pmatrix} \rho & 0 & 0 & 0 \\ 0 & p & 0 & 0 \\ 0 & 0 & p & 0 \\ 0 & 0 & 0 & p \end{pmatrix}\right)^2$$

2*(classic GR exact solution^2)

Superimposed tensors

Keep in mind that the point in squaring tensors in Complex EFE is to treat it like a superposition QM wavefunction Ψ*Ψ or (3+3i) (3-3i)=3²+3² to get a real value(Probability interpretation were it can collapse into one state). For the new Complex EFE it will have a new Scalar (2) throughout the field equation. The reason why the tensors can have the form below is that all the real parts are the same similar to (3+3i)(3-3i)=3^2+3^2= 2(3^2). In which we have a symmetry of equal parts real mass and imaginary mass. For the complex conjugate universe it to has equal parts real mass and imaginary mass. We live in a complex conjugate mass universe. The equation below captures all the energy from our complex conjugate pair big bang. As a result it transforms nonlinear GR into a linear GR equation that can give new exact solutions.

$$G_{2(\mu\nu^2)} = \left(\frac{128\pi^2(G^2)}{c^8}\right)T_{2(\mu\nu^2)}$$

2*(classic GR exact solution^2)

Starting with Vacuum solutions

Because were using a complex conjugate pair multiplication of the constants to get $(128\,\pi^2(G^2)/c^8)T_{2(uv^2)}$ the tensors and scalars must be treated the same way. Meaning the Complex GR field equation uses the same real parts for the tensors and scalars analogous to $(3+3i)(3-3i)=3^2+3^2=2(3^2)$ because were assuming the universe has equal parts real & imag. mass. The Complex EFE below is analogous to multiplying two complex conjugate points together to retrieve a real value like multiplying $\Psi^*\Psi$ to retrieve a real value.

$$R_{2(\mu\nu^2)} - \frac{1}{2}g_{2(\mu\nu^2)}(2R^2) = \left(\frac{128\pi^2(G^2)}{c^8}\right)T_{2(\mu\nu^2)}$$

$$G_{2(\mu\nu^2)} = \left(\frac{128\pi^2(G^2)}{c^8}\right)T_{2(\mu\nu^2)} \approx \psi^*\psi$$

Linear GR vacuum   NL GR vacuum

$$G_{2(\mu\nu^2)} = \left(\frac{128\pi^2(G^2)}{c^8}\right)T_{2(\mu\nu^2)} \qquad R_{\mu\nu} - \frac{1}{2}g_{\mu\nu}R = \frac{8\pi G}{c^4}T_{\mu\nu}$$

We need both L. GR and NL GR to describe classical phenomena and QG phenomena.

Linear vacuum solutions

It appears the new GR retains the fundamental properties of classical GR but just superimposed. The reason why I use (3+3i)(3-3i) as a example for the Complex EFE and not (3+5i)(3-5i) or (4+6i)(4-6i) is that we had a complex conjugate pair big bang with equal parts real & imaginary mass & it's complex conj. pair.

$$\left(\frac{8\pi G}{c^4} + \frac{8\pi G}{c^4}i \right)\left(\frac{8\pi G}{c^4} - \frac{8\pi G}{c^4}i \right) = \frac{128\pi^2(G^2)}{c^8}$$

$$G_{2(\mu\nu^2)} = \left(\frac{128\pi^2(G^2)}{c^8} \right) T_{2(\mu\nu^2)}$$

Because I use the same values to derive 128 π²(G²)/c⁸) the Tensors & Scalars in the Complex EFE must use same values for their real parts as well similar to (3+3i)(3-3i)=3^2+3^3=2(3^2).

We can take known solutions to Classical CR and expand them into Wavefunction GR form to study how they behaves as Linear GR. It's solutions describe Quantum Gravity effects.

2*(classic GR exact solution^2)

Compton wavelength for the Linear Graviton

The Compton wavelength is a quantum mechanical property of a particle.

$$\lambda = \frac{\hbar}{mc}$$

To learn more about the massive Graviton from the new Linear GR formula.

$$G_{2(\mu\nu^2)} = \left(\frac{128\pi^2(G^2)}{c^8}\right)T_{2(\mu\nu^2)}$$

The coupling constant or mass range of the Linear Graviton is conjectured to be

$$a = \pi^2/128$$

This alpha term was derived from the complex path integral in curved spacetime.

$$\int d^4x\sqrt{-((g+gi)(g-gi))}\left[L_{QFT} + \frac{\pi^2}{128}((R+Ri)(R-Ri))\right]$$

$$\alpha = (1/16\pi + 1/16\pi i)(1/16\pi - 1/16\pi i) = \frac{\pi^2}{128}$$

$a = \pi^2/128$ can reproduce the Complex EFE like $a = 1/16\pi$ can reproduce the normal EFE from the normal path integral.

<u>Compton wavelength for the linear Graviton</u>
To learn more about the massive Graviton
(a = $\pi^2/128$) we can study it's Compton
wavelength. It's important to note that the
Linear Graviton is related to a superimposed
wavefunction interpretation of the stress-energy tensor like $\Psi^*\Psi$. The Graviton associated
with linearized GR is the true Graviton.

<u>The new Linear Gravitational waves</u>
The linear version of GR structured like a
complex conjugate pair wavefunction $\Psi^*\Psi$

$$G_{2(\mu\nu^2)} = \left(\frac{128\pi^2(G^2)}{c^8}\right)T_{2(\mu\nu^2)}$$

describes a Graviton with different properties when compared to the classic nonlinear
GR Graviton. Consequently the Gravitational
wave associated with this new linear Graviton
will be different from regular Gravity waves.
Possibly we have <u>two types of Gravity waves</u>:
1. Classic GR non-linear Gravity waves
   from single eigenstate mass particles.
2. QG Linear GR Gravity waves from
   superimposed particles of matter
   from the complex pair big bang(WMAP)

Foundation for Linear GR QFT

Wavefunction GR(also called Linear GR) describes the Graviton as superimposed particles analogous to QM $\Psi^*\Psi$. The number "128" in LGR describes the new Graviton.

Linear GR	Non-linear GR
New Graviton	Graviton for NL GR

$$G_{2(\mu\nu^2)} = \left(\frac{128\pi^2(G^2)}{c^8}\right)T_{2(\mu\nu^2)} \qquad R_{\mu\nu} - \frac{1}{2}g_{\mu\nu}R = \frac{8\pi G}{c^4}T_{\mu\nu}$$

Determining the new G. rest mass

The # 128 in the new GR formula conjecturally point towards a Linear Graviton similar to the W-Boson 1/128 or EM 1/137. We can retrieve an alpha term related to Complex GR by using the path integral in complex curved space. The alpha term is π^2/128. If we use alpha 1/16π for a regular path integral in curved space then we can derive classic EFE. The value π^2/128 can derive linear EFE.

$$\int d^4x\sqrt{-((g+gi)(g-gi))}\left[L_{QFT} + \frac{\pi^2}{128}((R+Ri)(R-Ri))\right]$$

$$\alpha = (1/16\pi + 1/16\pi i)(1/16\pi - 1/16\pi i) = \frac{\pi^2}{128}$$

Rest mass is related to alpha term

The alpha term from the complex path integral is related to the coupling constant for the Linear Graviton. Value π^2/128 isn't describing a single massive Graviton but the superposition state of Massless Spin 2 Gravitons.

"G" alpha term constant

$$\alpha = \frac{\pi^2}{128}$$

W Boson

$$\alpha \approx 1/128$$

at $Q^2 \approx m^2(W)$

In the context of QFT for gravity we need the energy of π^2/128 to vibrate the field. The Quantum Field for gravity will only accept energy at this level or above.

Vibrating the Quantum Field for Gravity

At the coupling constant energy of π^2/128 or above the Linearized Graviton is formed. The quantum field for gravity is similar to other quantum fields. However, it's Feynman diagram describes particles as complex points. After the Quantum field is excited it forms a linear Graviton. Once form it diffuses into single eigenvalue Gravitons that correspond to each of the universes from the complex pair big bang (i.e. (3+3i)(3-3i)=2(3^2)).

187

How do the Linear Gravity Feynman diagram work?

The classical viewpoint of gravity exchanges is that we have 2 masses that exchanges massless Spin 2 Gravitons. However, Physicists found out that the model doesn't work. That gravity model is non-renormalizable. It produces infinities that can't be corrected. What Linear GR

$$G_{2(\mu\nu^2)} = (\frac{128\pi^2(G^2)}{c^8})T_{2(\mu\nu^2)}$$

tells us when QM is applied is the Graviton isn't a single point particle. It first exists in a superposition(Linear) form such as
i.e. (3+3i)(3-3i)=2(3^2).

A COMPLEX # FEYNMAN DIAGRAM

The point particle interactions for Linear GR is described by a complex point(a+bi) and it's complex conjugate point(a-bi) location. It describes a Graviton and it's antiparticle. Their annihilates makes the linear Graviton. Afterwards it diffuses back into single eigenstate Gravitons. It goes from 2(3^2) to (3+3i)(3-3i).

PATH INTEGRAL FEYNMAN DIAGRAM

The Feynman diagram for Linear Gravity describes single particles as complex point particles in a complex Quantum field plane.

$$\int d^4x \sqrt{-((g+gi)(g-gi))}[L_{QFT} + \frac{\pi^2}{128}((R+Ri)(R-Ri))]$$

$$\alpha = (1/16\pi + 1/16\pi i)(1/16\pi - 1/16\pi i) = \frac{\pi^2}{128}$$

COMPLEX POINT ON A SPRING

QFT is pictured as a mattress made of ball-springs. The QFT for gravity also have that setup except the ball is a complex point(a+bi) on a spring & it's describing a point location.
Oscillate up: 1/16π+1/16πi(complex point)
Oscillate down: 1/16π-1/16πi(comp. point)
The real part +1/16π is the stationary part and the imaginary part is oscillating between +1/16πi(gravity hill) and -1/16πi(gravity well) as oscillations in a Quantum Field of gravity.

189

Complex point exchanges

Complex values represent particle locations
similar to electron-positron annihilation.

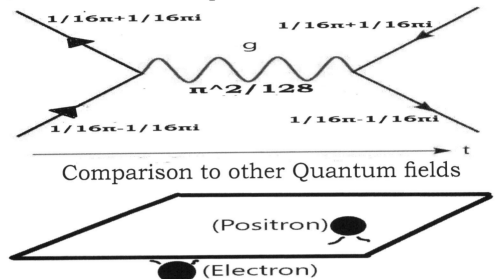

Comparison to other Quantum fields

Simply replace the electron point particle
with a complex point(i.e.(3+3i)) and replace
the positron particle with(i.e.(3-3i)). <u>The real
axis is the stationary vacuum</u>. The complex
values in the gravity diagram <u>represent single
complex points</u> similar to a electron,positron.

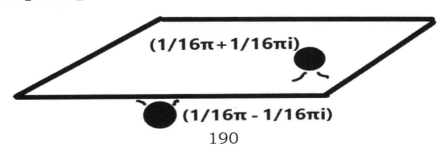

Complex point-spring QFT mattress

The QFT for gravity is based on a mattress spring analogy. However, we're replacing the ball-spring with a complex point ball-springs. Each ball is represented as a complex point in a complex Quantum Field.

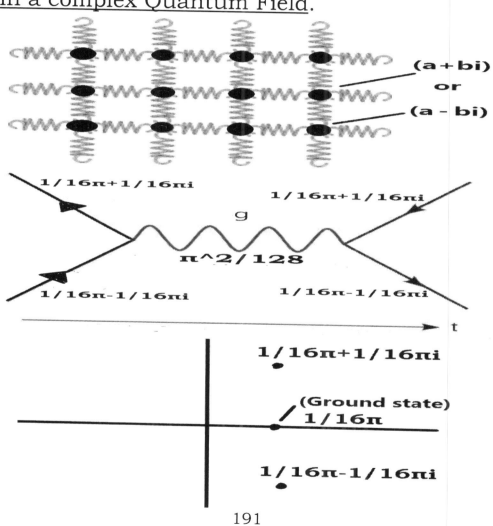

$1/16\pi + 1/16\pi i$

$1/16\pi + 1/16\pi i$

g

$\pi^2/128$

$1/16\pi - 1/16\pi i$

$1/16\pi - 1/16\pi i$

t

$1/16\pi + 1/16\pi i$

(Ground state)
$1/16\pi$

$1/16\pi - 1/16\pi i$

(a + bi)

or

(a - bi)

191

Complex point on a spring

A Ball-spring can also be expressed as a complex point on a spring. The two values(a+bi) is describing a single particle on a complex plane. When it oscillate it's a complex oscillation. The <u>real plane is the stationary vacuum</u>. The <u>imaginary plane is where the oscillations</u> and vacuum fluctuations take place like a up and down drum.

$$1/16\pi+1/16\pi i$$

(Ground state)
$$1/16\pi$$

$$1/16\pi-1/16\pi i$$

<u>Picture (3+3i), (3-3i) like a Electron-positron.</u>
The real axis above is the stationary vacuum state. The complex points represent a single particle and it's opposite particle.

The challenge with constructing a QFT for gravity is that we need a particle-anti particle annihilation that can make a Linearized Graviton. It doesn't appear positive matter(+) & positive mass antimatter(it has opposite charge) can make a Linear Graviton. Also if we try to collide real mass and negative mass it wouldn't annihilate and make a Graviton because of the "Runaway motion" effect. The negative mass and positive mass can never physically meet to annihilate. What other pair production pairs can be used other than particles-antiparticles or real mass-negative mass particles? The only other pair left is the complex point particle (a+bi) and it's complex conjugate point particle pair (a-bi). In which the complex value(a+bi) doesn't represent 2 particles but one particle located in a complex plane. The complex conj. value (a-bi) is analogous to it's antimatter particle point location. Their Tachyon gravity parts annihilate & make the linear Graviton (i.e. $(2+2i)(2-2i)=2(2^2)$). The L. Graviton can revert back into a separate complex value points(a+bi) &(a-bi). The particles are represented as <u>single complex point particles</u> on a Q. field of G.

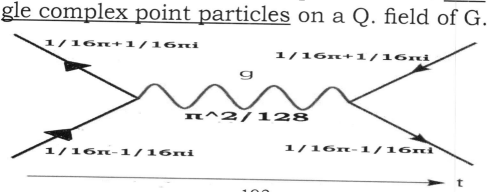

$1/16\pi+1/16\pi i$

$1/16\pi+1/16\pi i$

g

$\pi^2/128$

$1/16\pi-1/16\pi i$

$1/16\pi-1/16\pi i$

t

Picture them as complex points (a+bi) in a complex space Q. Field

The Linearized Graviton is based on linearized GR.

$$G_{2(\mu\nu^2)} = \left(\frac{128\pi^2(G^2)}{c^8}\right)T_{2(\mu\nu^2)}$$

If we express the path integral in complex conjugate pair like Linear GR then we derive an alpha term that describes the Graviton coupling energy constant.

$$\int d^4x \sqrt{-((g+gi)(g-gi))}\left[L_{QFT} + \frac{\pi^2}{128}((R+Ri)(R-Ri))\right]$$

$$\alpha = (1/16\pi + 1/16\pi i)(1/16\pi - 1/16\pi i) = \frac{\pi^2}{128}$$

The complex value (1/16π+1/16πi) is analogous to a complex point. The complex conjugate value (1/16π-1/16πi) is analogous to a complex conj. pair point. <u>The two complex points (1/16π+1/16πi) & (1/16π-1/16πi) below are similar to a electron and positron</u>.

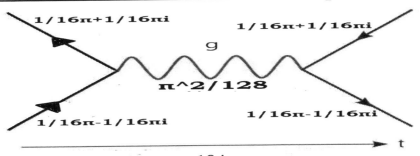

1/16π+1/16πi   1/16π+1/16πi

g

π^2/128

1/16π-1/16πi   1/16π-1/16πi

t

Explaining the Feynman digram for gravity

The Feynman diagram for gravity is based on a complex path integral. The alpha term related to Complex EFE describes Gravitons. The value $\pi^2/128$ is for the Graviton like $1/128$ for W Boson or $1/137$ for the EM force

$$\int d^4x \sqrt{-((g+gi)(g-gi))}[L_{QFT} + \frac{\pi^2}{128}((R+Ri)(R-Ri))]$$

$$\alpha = (1/16\pi + 1/16\pi i)(1/16\pi - 1/16\pi i) = \frac{\pi^2}{128}$$

1. $(1/16\pi + 1/16\pi i)$ is a <u>nonlinear Graviton</u> particle in a complex spacetime plane.
2. $(1/16\pi - 1/16\pi i)$ is complex conjugate g.

<u>Picture the F.D. as complex points on a complex Q. Field plane</u>

In the Feynman diagram below imagine the complex values as complex points. <u>Each complex point is analogous to a single particle.</u>

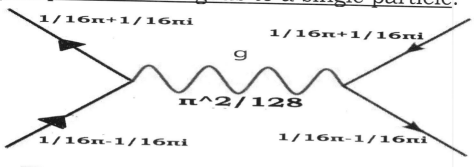

195

Linear Covariant quantization

Traditionally the metric is divided into a "background" g_{uv} and a quantum fluctuation h_{uv}.

$$g_{\mu\nu} = \bar{g}_{\mu\nu} + \sqrt{16\pi G}\,h_{\mu\nu}$$

Just as Linear GR is based on the FOIL method (a+bi)(a-bi) we can do the same for the equation above. For example, one of the terms is expressed based on the FOIL method.

$$\left(\sqrt{16\pi G} + \sqrt{16\pi G}\,i\right)\left(\sqrt{16\pi G} - \sqrt{16\pi G}\,i\right)$$

We can take the entire equation and express as complex conjugate pairs(i.e. (3+3i)(3-3i)=2(3^2)) analogous to complex wavefunction multiplication Ψ*Ψ.

$$g_{2(\mu\nu^2)} = \bar{g}_{2(\mu\nu^2)} + 32\pi G h_{2(\mu\nu^2)}$$

A superimposed background metric and it's quantum fluctuation

The complex point virtual particles of a superimposed background metric are related to a Quantum field of Linear Gravity.

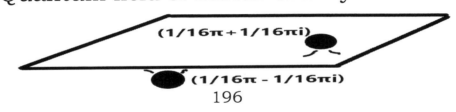

(1/16π + 1/16πi)

(1/16π - 1/16πi)

Gravity virtual particles

A complex space path integral is used to construct a Feynman diagram for the Graviton.

$$\int d^4x \sqrt{-((g+gi)(g-gi))}[L_{QFT} + \frac{\pi^2}{128}((R+Ri)(R-Ri))]$$

If the alpha term is $1/16\pi$ for a regular path integral in curved spacetime then we can derive the classic EFE. If we expand the path integral in complex conjugate pair curved space time then we have the alpha term as

$$\alpha = (1/16\pi + 1/16\pi i)(1/16\pi - 1/16\pi i) = \frac{\pi^2}{128}$$

were in principle can derive the complex linear EFE. Think of the complex values above as the complex point location of particles in a Feynman diagram of Quantum gravity(complex valued spacetime curvature GR model).

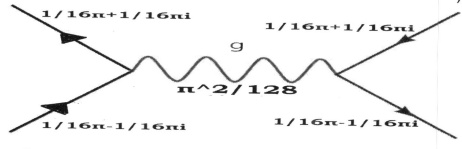

Gravity/antigravity annihilate to make Linear Gravitons $\pi^2/128$ like Ele./Pos. annihilation.

Laying the groundwork for gravity QFT

If the Quantum Field theory for gravity is
based on a complex plane

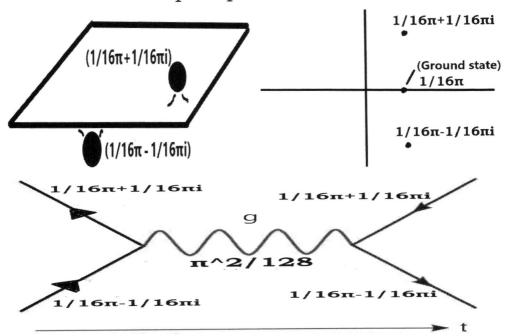

then it may be possible to extend the complex
nature of linear GR based on (a+bi)(a-bi)=
a^2+b^2 like linear Ψ*Ψ to QFT.

$$R_{(\mu\nu)^2+(\mu\nu)^2} - \frac{1}{2}g_{(\mu\nu)^2+(\mu\nu)^2}((R+Ri)(R-Ri)) = ((\frac{8\pi G}{c^4})^2 + (\frac{8\pi G}{c^4})^2)T_{(\mu\nu)^2+(\mu\nu)^2}$$

The complex nature of GR is the ground work
for constructing a gravity QFT.

Groundwork for Gravity QFT

QFT is based on complex numbers. For example, if the action S is real, then e^{iS} is a unit-magnitude complex # on the unit circle in the complex plane(x+yi).

$$e^{iS} = \cos S + i \sin S$$

Just as linear GR is based on complex conjugate pair multiplication

$$G_{2(\mu\nu^2)} = \left(\frac{128\pi^2(G^2)}{c^8}\right)T_{2(\mu\nu^2)}$$

we can extend that complex nature to a QFT for gravity. General Relativity can use
$$\text{Cos} + \text{Sin}$$
in the area of gravity waves. We have complex gravity waves based on complex linear GR.

1. Real gravity wave + imag. gravity wave

2. Complex conjugate pair gravity wave
The real part is the ground state. I haven't work out the rigorous details but it may be possible to make QFT formulation for gravity based on Linear GR.

A new LINEAR Einstein-Hilbert action

The Einstein-Hilbert action is expressed as follows in spacetime with no matter fields.

$$S = \frac{1}{16\pi} \int d^4x \sqrt{-g} R + S_m[\psi, g_{\mu\nu}]$$

Let's expressed the E-H action like a Complex conjugate pair wave function were you multiply a wavefunction pair to retrieve a real value analogous to (a+bi)(a-bi)=a^2+b^2.(i.e. (3+3i)(3-3i)= 3^2+3^2= 2(3^2)).

$$S = \frac{\pi^2 G^2}{128} \int d^4x \sqrt{-(2(g^2))} 2R^2 + 2(S_m[\psi, g_{\mu\nu}])^2$$

What is so exciting is the term 1/128 in the wavefunction like Einstein-Hilbert action. Remember the energy of the W-Boson is 1/128.

$$\alpha \approx 1/128$$

$$\text{at } Q^2 \approx m^2(W)$$

For this QM description of GR as a wavefunction we see "128" like 1/128 for the W-Boson!

The number 128 in the LINEAR E-H action

When we treat the Einstein Field Equation like a complex conjugate pair wave function similar to Ψ*Ψ based on (a+bi)(a-bi)=a^2+b^2 it yields a number that is similar to the alpha value for the W-Boson. In other words this Quantum theory of GR contains the number "128" like 128 in the W-Boson(1/128). The following highlights the three wavefunction like Einstein Field Equations that yields the term 128 like the W-boson 1/128. Notice "128" in the Eq. below.

$$R_{2(\mu\nu^2)} - \frac{1}{2}g_{2(\mu\nu^2)}(2R^2) = (\frac{128\pi^2(G^2)}{c^8})T_{2(\mu\nu^2)}$$

The following is a complex conj. pair path integral in curve spacetime similar to (a+bi)(a-bi)=a²+b².(1/16π+1/16πi)(1/16π-1/16πi) =π²/128 can derive the Linear GR EFE.

$$S_{QFT+gravity} = \int d^4x\sqrt{-(2g^2)}[L_{QFT} + \frac{\pi^2}{128}2R^2]$$

A New Complex E-H action is as follows:

$$S = \frac{\pi^2G^2}{128}\int d^4x\sqrt{-(2(g^2))}2R^2 + 2(S_m[\psi, g_{\mu\nu}])^2$$

Notice the # "128" like the W-Boson 1/128.

201

Another LINEAR Einstein-Hilbert action

In a slightly different formulation of the Einstein-Hilbert action we see another possible connection to real world phenomena. We take the E-H and multiply: (a+bi)(a-bi)=a^2+b^2.

$$S_{EH} = \frac{c^4}{16\pi G} \int d^4x \sqrt{-g} R$$

$$S_{EH} = \frac{c^8}{128\pi^2 G^2} \int d^4x \sqrt{-(2(g^2))} 2R^2$$

We took the complex conjugates of E-H including the constants. We have $c^8/128\pi^2 G^2$

$$\left(\frac{1}{2 \times \frac{8\pi G}{c^4}} + \frac{1}{2 \times \frac{8\pi G}{c^4}} i \right) \left(\frac{1}{2 \times \frac{8\pi G}{c^4}} - \frac{1}{2 \times \frac{8\pi G}{c^4}} i \right)$$

The complex conjugate pair multiplied E-H above is related to the Complex EFE.

$$G_{2(\mu\nu^2)} = \left(\frac{128\pi^2(G^2)}{c^8} \right) T_{2(\mu\nu^2)} \qquad S_{EH} = \frac{c^8}{128\pi^2 G^2} \int d^4x \sqrt{-(2(g^2))} 2R^2$$

What's interesting is the constant $128\pi^2 G^2/c^8$ in the Complex EFE is inverted into $c^8/128\pi^2 G^2$ for the Complex E-H action. The 2nd interesting thing is that $c^8/128\pi^2 G^2=1.159$ x 10^85 is a huge #. The # of atoms in our universe is est. 10^78-10^86.

The LINEAR E-H action & Quantum Gravity

$$S = \frac{1}{16\pi} \int d^4x \sqrt{-g} R + S_m[\psi, g_{\mu\nu}]$$

Let's take the complex conjugate pair form similar to (3+3i)(3-3i)=2(3^2) like multiplying two wavefuntions together: Ψ*Ψ.

$$S = \frac{\pi^2 G^2}{128} \int d^4x \sqrt{-(2(g^2))} 2R^2 + 2(S_m[\psi, g_{\mu\nu}])^2$$

Notice the first term $\pi^2 G^2/128$. If we break it up then it may describe the energy coupling constant of the Graviton $\pi^2 G^2/128$ multiplied by G^2. Remember the term $\pi^2 G^2/128$ is similar to the alpha term coupling constant for the W-Boson.

$$\alpha = \frac{\pi^2}{128} \qquad \alpha \approx 1/128$$

at $Q^2 \approx m^2(W)$

The $\pi^2 G^2/128$ is describing the Graviton coupling constant value times G^2.

A new Gravity alpha term from Complex EFE

Path integral in curved spacetime is a follows:

$$S_{QFT+gravity} = \int d^4x \sqrt{-g}[L_{QFT} + \alpha R]$$

where we could derive $\alpha = 1/16\pi$ to reproduce the Einstein Field Equations. We can
treat the above equation like a complex conjugate pair wave function $\Psi^*\Psi$ to retrieve
real values similar to $(3+3i)(3-3i)=2(3^2)$. The
Complex path integral in curved space will
have the same real values for $(a+bi)(a+bi)$.

$$\int d^4x \sqrt{-((g+gi)(g-gi))}[L_{QFT} + \frac{\pi^2}{128}((R+Ri)(R-Ri))]$$

Just as $1/16\pi$ can reproduce the normal EFE
for a regular path integral in spacetime, we
can derive $\alpha = \pi^2/128$ to reproduce the complex conjugate pair EFE.

$$\alpha = (1/16\pi + 1/16\pi i)(1/16\pi - 1/16\pi i) = \frac{\pi^2}{128}$$

What's amazing is that this new alpha term
$\alpha = \pi^2/128$ in complex conjugate curved
space is similar to the alpha term in the value
of a W Boson!

$$\alpha \approx 1/128$$

$$\text{at } Q^2 \approx m^2(W)$$

204

Is "α=π^2/128" describing a new particle?
(Describing a Massive Graviton)

Is the alpha term "α = π^2/128" from the Complex path integral in curved space time describing a new gravity related particle? It's value is similar to that of the W Boson were α =1/128. This may be the <u>true Graviton</u>!

If α =π^2/128 = 0.07710628438 is a new discovery term then where else in physics do we see the term? Surprisingly this same term π^2/128 = 0.07710628438 shows up in a Quantum Mechanical particle in a Box exercise. <u>Do a Google search and type in</u>

0.07710628438

The Google search will show a Arvix paper. Go to part 3 in the Arvix paper and you will see the term α =π^2/128 = 0.07710628438 related to a QM particle in a box. This section shows hypothetically if the EFE are treated like a complex conjugate pair analogous to Ψ*Ψ then it yields out a constant term α =π^2/128 that is similar to the W-Boson α =1/128 and a QM particle in a box! Complex GR has pressure fluctuations from it's tensor & is related to α =π^2/128 in the Arvix paper.

Making Linear GR UV complete

What's interesting about the coupling constant

$$\alpha = \pi^2/128$$

from the complex curved space path integral

$$\int d^4x \sqrt{-((g+gi)(g-gi))}\left[L_{QFT} + \frac{\pi^2}{128}((R+Ri)(R-Ri))\right]$$

is that it also shows up in a arvix paper re-
garding a QM particle in a box. That particular QM in a box describes a system that is
analogous Feynman diagram loops. What's so
interesting between the alpha term
$\alpha = \pi^2/128$ from the Complex path integral
and the alpha term $\alpha = \pi^2/128$ in the arvix
paper isn't just their numerical correlation
"$\pi^2/128$" but the fact that both are related
to Feynman diagrams. The arvix paper describes loop topology Feynman diagrams and
when it comes to quantum gravity we can
have the issue of Feynman loops in the context of ultraviolet divergences.

Do a Google search and type in
0.07710628438

The Arvix paper has the same alpha term
$\alpha = \pi^2/128 = 0.07710628438$ as the complex integral in curved spacetime.

$$\int d^4x \sqrt{-((g+gi)(g-gi))}\left[L_{QFT} + \frac{\pi^2}{128}((R+Ri)(R-Ri))\right]$$

The pressure fluctuation in the Einstein stress-energy Tensor in the context of the Linear EFE below

$$G_{2(\mu\nu^2)} = \left(\frac{128\pi^2(G^2)}{c^8}\right)T_{2(\mu\nu^2)}$$

may relate to the pressure fluctuations from the arvix paper. Both the Arvix paper and linear GR relates to Feynman diagrams. This Feynman diagram topology connection could help us determine how to make Linear GR UV complete!

The Linear GR fine structure constant
(A new Gravity coupling constant)

When we treat EFE as a complex conjugate pair wavefunction we derive what appears to be a GR fine structure constant. We multiply EFE as complex conjugate pairs(i.e. (3+3i)(3-3i)=2(3^2)) to retrieve new real values similar to $\Psi^*\Psi$ being multiplied together.

$$\int d^4x \sqrt{-((g+gi)(g-gi))}\left[L_{QFT} + \frac{\pi^2}{128}((R+Ri)(R-Ri))\right]$$

The focus for now isn't on solving the complex EFE because it's analogous to a smeared Complex wave function $\Psi^*\Psi$. The GR expression above is analogous to a probability distribution. The number π^2/128 is related to the Gravity coupling constant between gravity and it's complex conjugate pair gravity similar to the value of the W Boson at 1/128. Both are constants and both have "128" in common. "128" is the magic number.

"G" alpha term constant   W Boson

$$\alpha = \frac{\pi^2}{128}$$ $\alpha \approx 1/128$

at $Q^2 \approx m^2(W)$

The C.curved Path integral is treated like a Ψ.

$$\int_T \psi^* \psi \, dT = 1$$

We can multiply complex conjugates to retrieve a new real value like (3+3i)(3-3i)=2(3^2)

$$\int d^4x \sqrt{-((g+gi)(g-gi))}\left[L_{QFT} + \frac{\pi^2}{128}((R+Ri)(R-Ri))\right]$$

The equation above is based on the following.

$$\alpha = (1/16\pi + 1/16\pi i)(1/16\pi - 1/16\pi i) = \frac{\pi^2}{128}$$

We could derive Complex EFE. The alpha
term from the path integral in complex spacetime is similar to the W-Boson 1/128 were
π^2/128 may be describing a new superposition of massless spin 2 Gravitons.

Complex spacetime path integral

The path integral in complex spacetime (a+bi)
(a-bi) reveals the value of the linear Graviton. The new gravity related alpha term α
= π^2/128 is special because it's similar to
the α=1/128 for the W boson or the EM force
1/137. This formulation of the path integral
in complex spacetime is treating it similar to
a QM wavefunction were two wavefunctions
Ψ*Ψ multiply to give real values.

The path integral in Complex spacetime

The premises of Wavefunction GR is that it's not a literal wavefunction form of GR but an analogous form of QM wavefunction mechanics. We multiply parts of EFE like a complex conjugate pair wave-function (a+bi)(a-bi) Ψ*Ψ.

$$\Psi^*(x,t)\Psi(x,t)$$

A path integral in complex spacetime can be expressed as a complex conjugate pair analogous to Ψ*Ψ.

$$\int d^4x \sqrt{-((g+gi)(g-gi))}\left[L_{QFT} + \frac{\pi^2}{128}((R+Ri)(R-Ri))\right]$$

The following alpha term can derive the complex EFE below.

$$\alpha = (1/16\pi + 1/16\pi i)(1/16\pi - 1/16\pi i) = \frac{\pi^2}{128}$$

$$G_{2(\mu\nu^2)} = \left(\frac{128\pi^2(G^2)}{c^8}\right)T_{2(\mu\nu^2)}$$

For QM wavefunctions the sum of the probabilities for all of space must be equal to 1.

$$\int \Psi^* \Psi \, dr = 1$$

To normalize the path integral for the complex spacetime: (a+bi)(a-bi) we can also have it equal to one.

LINEAR SPACTIME PATH INTEGRALS

$$S_{QFT+gravity} = \int d^4x \sqrt{-(2(g^2))} \left[L_{QFT} + \frac{\pi^2}{128} 2R^2 \right] = 1$$

$$S = \int d^4x \sqrt{-(2(g^2))} \, R_{2(\mu\nu^2)} = 1$$

Space Path integral Probability equal to 1

It's to be determined how to cause the GR related path integrals to equal 1. If those GR related probility equations don't equal to 1 then it's the same as a QM wavefunction not equaling to one. Meaning the wavefunction won't be normalized.

We need the sum of "spacetime path integrals" equal to 1 in order for it to be compatible with QM were it's probabilistic wavefunction equations equal 1.

ADM formalism for Linear GR(LGR)

(A Wheeler–DeWitt type harmonic oscillator)
The graph on the left represent inflationary
and deflationary curves of the universe based
on a complex conjugate pair (a+bi)(a-bi) big
bang and galaxy rotational curves. We can
correlate the cosmological and galaxy rotational curves to QM harmonic oscillators.

<u>Gravity curves</u>   <u>QM harmonic oscillators</u>

<u>Imag matter(+,- x^2)</u>   <u>QM Oscillator +,- x^2</u>
<u>Real matter (+,- $1/x^2$)</u>   QM Oscillator +,- $1/x^2$
Based on the correlation above we can draft a
ADM formalism for linear GR were spacetime
(Gravity rotational/universe curves) is foliated like QM harmonic oscillator waveforms.
This ADM formalistic harmonic oscillator is
scaled up to galaxy rotation or universe scale
with billions or trillions of waveforms for a
spatial slice level analogous to E10^60.

ADM formalism for Linear GR(LGR)

(A Wheeler–DeWitt type harmonic oscillator)
Ingredients for Linear GR ADM formalism:
1. The ADM spatial metric slices for Linear GR are analogous to the discrete wave functions of a Schrodinger harmonic oscillator.
2. Each spatial slice is represented as a integer number analogous to the Schrodinger QM harmonic oscillator energy levels $E1, E2, E3....$
3. This ADM formalism for a spatial harmonic oscillator describes gravity rotational curves and universe inflationary-deflationary curves.

Linear GR ADM formalism isn't directly related to GR but the gravity rotational curves & the expansion-contraction of our universe related to gravity. Just as we have a Schrodinger equation for a QM harmonic oscillator we can have a new type of Schrodinger equation for a ADM metric harmonic oscillator. Meaning space is foliated like a QM harmonic oscillator and the metric is oscillating like a wave at integer values. The ADM spatial harmonic oscillator describes galaxy rotational curves & the universe inflationary-deflationary curves.

213

REVIEW OF LINEAR GR(LGR)

Conjecturally, classical GR is incomplete in the context it's not capturing all the energy in the Stress-energy tensor. For instance, if you look at a QM wavefunction Ψ*Ψ it consists of 4 parts (a+bi)(a-bi). Likewise, in a QM interpretation of GR express each part of a complex wavefunction(a+bi)(a-bi) as a separate Stress-energy tensor. The 4 different Stress-energy tensors in Wavefunction GR are analogous to the 4 terms in a wavefunction (a+bi)(a-bi). Also these 4 different parts of the Stress-energy tensors are based on a complex conjugate pair big bang universe $E=2c^4m^2 = (mc^2+(mc^2)i)(mc^2-(mc^2)i)$ & not simply a matter-antimatter ($E=mc^2$) big bang. Tensors & constants are expressed (a+bi)(a-bi) or i.e. (3+3i)(3-3i)=2(3^2) and their real parts are the same for each term. The following is EFE in the context of QM Ψ*Ψ. It has a complex conjugate pair Tensor like QM Ψ*Ψ.

$$G_{2(\mu\nu^2)} = (\frac{128\pi^2(G^2)}{c^8})T_{2(\mu\nu^2)} \approx \psi^*\psi$$

$$R_{2(\mu\nu^2)} - \frac{1}{2}g_{2(\mu\nu^2)}(2R^2) = (\frac{128\pi^2(G^2)}{c^8})T_{2(\mu\nu^2)}$$

214

Where are the linear Gravitons now?

The big bang was a linear complex conj. superimposed start. Associated with that bang was linear Gravitons. As the universe expanded the Linear Gravitons transitioned into single eigenstates(Massless Spin 2) Gravitons.

L. Gravitons scatter like W Bosons

Linear Gravitons(superpositioned Gravitons) have a high energy coupling constant. The single eigenstate Gravitons don't have mass but as Linear Gravitons they have high energy. That high energy equals a high mass similar to the W Boson. Just as the Higgs field helped the W Boson scattering behavior at high energy the Higgs field can also help the Linear Graviton scattering at high energies.

<u>Linear Graviton coupling</u>   <u>W Boson</u>

$$\alpha = \frac{\pi^2}{128} \qquad \alpha \approx 1/128$$

$$\text{at } Q^2 \approx m^2(W)$$

QFT theory for Gravity

A QFT for linearized GR can be constructed based on a complex point ball-spring mattress analogy. This complex field can make linearized Gravitons when excited enough.

The Primordial Linear Gravitons & Primordial linear Gravity waves

If Linear GR is correct then under what conditions do we see "Linear Gravitons" and "linear Gravity waves". According to Linear GR it's the Tachyons that is responsible for the expansion and acceleration of our universe.

If we had a complex big bang(a+bi) and a complex conjugate pair rather than matter-antimatter pair creation then it's the tachyon matter that is causing the expansion. Just as FTL Special Relativity physics works in reverse of our slower than light physics so do it's gravity. Just as regular gravity decreases over distance like $1/x^2$, the gravity for tachyons would work in reverse and have a x^2. Over time it accelerates in repulsion. If Tachyons from a complex big bang are causing our universe to expand and accelerate then what role do "linear Gravitons" play in our current universe. The answer is none. Gravitons started out linear at the big bang.

Linear Gravitons at Big bang

The conjecture is that we had a complex conjugate pair big bang (a+bi)(a-bi) analogous to a superposition wavefunction $\Psi^*\Psi$. Furthermore the Graviton was superimposed (linear) as well. However, as the universe grew apart it diffused from it's superimposed linear form and into its nonlinear form. Correspondingly, each of the 4 parts of the universe(a+bi)(a-bi) has its own single eigenstate Gravitons from the original superposition linear Gravitons. The Linear Gravitons were still massless but existed in a superposition massless state before collapsing into separate eigenstates.

Experimental proposals

1. Particle detection: In terms of a GUT it would be the linear Graviton that made up part of the Superforce. This linear Graviton coupling constant is conjectured to be $\pi^2/128$ similar to W Boson coupling constant $1/128$.
2. Search for linearized G waves in the WMAP from a Complex conjugate pair Big bang.

Experimental Test 1 (Discover a new type of Linear Graviton Particle).

The new GR formulation below describes a new superimposed Linear Graviton based on the superimposed Linear GR formula below.

$$G_{2(\mu\nu^2)} = \left(\frac{128\pi^2(G^2)}{c^8} \right) T_{2(\mu\nu^2)}$$

The conjecture is that the new linear Graviton is related to the new EFE and is similar in mass(energy) to the W-Boson(1/128).

$$\alpha = \frac{\pi^2}{128} \qquad \begin{array}{c} \alpha \approx 1/128 \\ \text{at } Q^2 \approx m^2(W) \end{array}$$

The 128 in the new GR formula suggests it's describing a massive(superimposed) Graviton similar to the W-Boson. Coupling constant for the new Graviton is "$\pi^2/128$".

The linear Graviton

Classic GR Spin 2 massless Gravitons test have failed because we have being searching for a non-linear GR Graviton. The true Graviton is based on Linear GR.

Predicting a new LINEAR Graviton

When we look at the new GR formulation based on (a+bi)(a-bi) similar to Ψ*Ψ we see the number "128" like the 1/128 in the W-Boson. That could suggests that "alpha" related to the Linear Graviton is similar to the W-Boson's alpha ≈ 1/128 at Q^2 ≈ m^2(W). The stress-energy tensor below is describing a Linear Graviton that includes all possible forms of energy for the Stress-energy tensor in the context of QM wavefunctions(i.e. Ψ*Ψ).

$$G_{2(\mu\nu^2)} = \left(\frac{128\pi^2(G^2)}{c^8}\right)T_{2(\mu\nu^2)}$$

Properties of the new Linear Graviton

The Spin and behavior of the Graviton will come from the Stress-energy tensor. What stands out is the number "128" were that number appears to be related to the coupling constant energy. *At that energy level a particle detector may detect a linearized Graviton.*

"G" alpha term constant   W Boson

$$\alpha \;=\; \frac{\pi^2}{128}$$ $\alpha \approx 1/128$

at $Q^2 \approx m^2(W)$

The Geometric coupling constant value!

The Linear Graviton coupling "$\pi^2/128$ " is interesting because it's related to geometry(**π**).

Strong	a_s	1
Gravity	a_g	$\pi^2/128$ = **0.077106...**
Weak	a_w	1/128 = 0.0078125...
EM	a	1/137 = 0.007299...

The energy associated with $\pi^2/128$ = 0.077106... is massive. It's slightly less than the Nuclear strong force!

How the Gravity alpha term was determined

The challenge was to determine what is the alpha term for the new GR EFE as it relates to 128. One approach was to look for a alpha term that is related to the Path integral in complex conjugate pair curved space.

$$\int d^4x \sqrt{-((g+gi)(g-gi))} \left[L_{QFT} + \frac{\pi^2}{128}((R+Ri)(R-Ri)) \right]$$

$$\alpha = (1/16\pi + 1/16\pi i)(1/16\pi - 1/16\pi i) = \frac{\pi^2}{128}$$

The alpha term "$\pi^2/128$" can derive the Complex conjugate pair EFE like $1/16\pi$ can derive the classic EFE. The value "$\pi^2/128$" is the coupling energy of the new Graviton.

A new type of Supersymmetry
(Complex conjugate pair symmetry)

Gravity based on new Linear GR is slightly weaker than the Strong force but mightier than the Weak force and EM force.

$$G_{2(\mu\nu^2)} = \left(\frac{128\pi^2(G'^2)}{c^8}\right)T_{2(\mu\nu^2)}$$

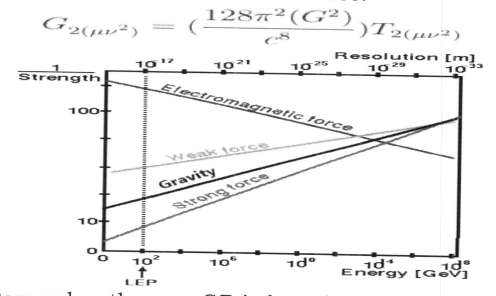

Remember the new GR is based on (a+bi) (a-bi) analogous to wavefunction multiplication $\Psi^*\Psi$. This new GR has a complex conjugate pair big bang universe. We have a mirror symmetry between our universe and it's complex conjugate pair similar to (3+3i) & (3-3i). Real particles are their own complex conjugate. Also the mirror symmetry universe may give a type of symmetry to the Weak force.

The new standard model

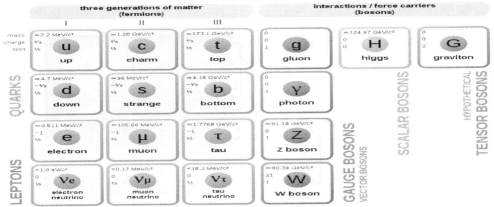

Standard Model of Elementary Particles and Gravity

Linear Graviton Boson

(It's not a literal Massive Graviton)

At the complex conjugate big bang the Graviton was linearized(superimposed). As the universe diffused the Linear Gravitons went into their separate Spin 2 massless states associated with NL GR(Single eigenstate Gravitons).

$$G_{2(\mu\nu^2)} = (\frac{128\pi^2(G^2)}{c^8})T_{2(\mu\nu^2)}$$

The coupling constant of "$\pi^2/128$" of the Linear Graviton based on Linear GR is similar to the coupling constant W Boson at $1/128$. However, the coupling constant isn't referring to a single Massive Graviton but superposition massless Spin 2 Gravitons.

Quantum field theory for Gravity

The Graviton was derived from a complex path integral. (1/16π + 1/16πi) is a <u>NL Graviton</u> complex point particle & (1/16π - 1/16πi) is it's NL antiGraviton. They annihilate to make a L. Graviton analogous to a electron & positron annihilating to make a photon.

$$\int d^4x \sqrt{-((g+gi)(g-gi))} [L_{QFT} + \frac{\pi^2}{128}((R+Ri)(R-Ri))]$$

$$\alpha = (1/16\pi + 1/16\pi i)(1/16\pi - 1/16\pi i) = \frac{\pi^2}{128}$$

What about the Higgs field for imaginary mass particles?

LGR is based on a complex conjugate pair big bang. If Higgs gives real mass it's mass then their should be a process for imag. mass. I don't know the full details but imaginary mass(+bi) has a real part. Conjecturally, the Higgs could give mass to it's real part & the imag. component gives it FTL properties.

223

Test 2: <u>Testing the complex conjugate pair big bang model from Primordial Gravity waves</u>

(LINEAR GRAVITY WAVE TEST)

The complex conjugate pair stress-energy tensor and geometric tensors are analogous to complex conjugate pair QM wave functions. Both uses the FOIL method to derive real value answers. The superimposed stress-energy tensor and superimposed geometry are analogous to a superposition QM wavefunction $\Psi^*\Psi$. This complex GR model is based on a complex conjugate pair big bang. The massive Graviton based on the new EFE is described as a superimposed Graviton analogous to a superposition wavefunction. Rather than having point like exchanges, we have superimposed wave-like Graviton exchanges. WGR is based on a complex conjugate pair big bang. If that existed then it should have generated gravity waves. We can compute complex conjugate pair Gravity waves from a big bang to see if they match the WMAP.

$$2\Box^2 2h^{\mu\nu^2} = 2\left(-\frac{\partial^2}{\partial t^2} + \nabla^2\right) 2h^{\mu\nu^2} = -\frac{512\pi^2 G^2}{c^8} 2T^{\mu\nu 2}$$

224

Linearized Gravity waves will be associated with Linear GR waves and explain observations that Nonlinear GR-Gravity waves can't.

$$G_{2(\mu\nu^2)} = \left(\frac{128\pi^2(G^2)}{c^8}\right)T_{2(\mu\nu^2)}$$

We can start with a Gravitational wave equation & express it like complex conjugate pairs (a+bi)(a-bi)=a^2+b^2 or (3+3i)(3-3i)= 2(3^2) similar to Ψ*Ψ.

$$\Box h^{\mu\nu} = \left(-\frac{\partial^2}{\partial t^2} + \nabla^2\right)h^{\mu\nu} = -\frac{16\pi G}{c^4}T^{\mu\nu}$$

$$2\Box^2 2h^{\mu\nu^2} = 2\left(-\frac{\partial^2}{\partial t^2} + \nabla^2\right)^2 2h^{\mu\nu^2} = -\frac{512\pi^2 G^2}{c^8}2T^{\mu\nu 2}$$

2*(classic GR exact solution^2)

Linear Gravity waves from a Complex conjugate pair big bang

If Linear GR is correct then we had a complex conjugate pair (a+bi)(a-bi) big bang of complex mass and not simple matter-antimatter pair creation. As a result the complex conjugate pair big bang should produce Linear Gravity waves that we can observe in WMAP.

Linear Gravity waves from a Complex conjugate pair big bang

The following is the new Gravity wave equation based on Complex conjugate pair multiplication(ie (3+3i)(3-3i)=2(3²) of Gravity waves. Real parts are the same because it's assumed we have equal amounts real & imag. mass.

$$2\Box^2 2h^{\mu\nu^2} = 2\left(-\frac{\partial^2}{\partial t^2} + \nabla^2\right)^2 2h^{\mu\nu^2} = -\frac{512\pi^2 G^2}{c^8} 2T^{\mu\nu 2}$$

The new Gravity wave equation based on (a+bi)(a-bi) is similar to taking 4 different EM waves and combining them into one overall EM wave. EM waves emitted from charged Tachyons should still travel at the speed of light like normal mass charged particles emit EM waves at the speed of light. Each of the 4 charged particles (2 real charged, 2 Tachyon charged particles) emit EM waves at the speed of light. Now, imagine overlapping all 4 EM waves from the moving 4 charge particles into one overall EM wave. This overlap EM wave is similar to the overlapped Gravity wave from the equation above. These Gravity waves could be Primordial gravity waves & evidence of a C. conjugate pair big bang.

226

New Primordial linear gravitational waves
(Complex conjugate pair big bang)

What the new wavefunction GR(WGR) model

$$G_{2(\mu\nu^2)} = (\frac{128\pi^2(G^2)}{c^8})T_{2(\mu\nu^2)}$$

suggests is that we had a complex conjugate pair big bang (a+bi)(a-bi) rather than just a matter and antimatter big bang. Just after the big bang we had 4 different gravitational wave sources based on (a+bi)(a-bi). Those 4 gravitational waves superimposed and overlapped to make one new combined Primordial gravitational wave signature.

$$2\Box^2 2h^{\mu\nu^2} = 2\left(-\frac{\partial^2}{\partial t^2} + \nabla^2\right)^2 2h^{\mu\nu^2} = -\frac{512\pi^2 G^2}{c^8} 2T^{\mu\nu^2}$$

Conjecturally, if we can observe these waves in the cosmic microwave background radiation then it would further support WGR.

We can <u>also</u> look for the <u>specific Gravity wave</u> <u>signature from it's collapse</u> into one G wave.

Superposition big bang and big crunch

Classical GR describes the big bang starting as a infinitely dense hot point. However, for the linearized version of GR it describes the beginning as a superposition starting point. For example, let's look at the equation and solution.   (3+3i)(3-3i)=2(3^2)

Each number "3" represent a different mass(energy) part of the earlier complex universe. Meaning we had a superimposed Mass(energy) state analogous to a superim-posed(superposition) QM wavefunction.

$$\Psi*\Psi$$

It's possible that the WMAP wasn't by Quantum fluctuations but from a superimposed energy(mass) state from linear Gravity! As the superimposed universe grew apart the universe went from a superposition state (i.e. 2(3^2)) into single eigenstates (3+3i)(3-3i).

We had a superposition Linear big bang and will have a superposition linear big crunch.

The fate of our complex pair universe

From our complex conjugate pair superimposed linear big bang (i.e. $(3+3i)(3-3i)=2(3^2)$ produce the following universe curves.

1. The real mass (3 and 3) has decreasing deflationary curves $(+/-\ 1/x^2)$.

2. The imag. mass(3i and -3i) is responsible for the expansion and acceleration of our universe $(+/-\ x^2$ curves). FTL Tachyon physics works in reverse of real matter physics.

3. Because imaginary energy lose energy as it accelerates their will be a point were it will lose so much energy that it's weaker than the pull of gravity. That will cause the universe & it's pair to each collapse into big crunches.

Superposition Primordial Gravity waves

Imagine gravity waves associated with each real part of the complex conjugate pair big bang(a+bi)(a-bi). Superimposed Gravity waves in the WMAP will prove a linear big bang.

Linear E-H action & it's link to Gravity waves

Let's explore a E-H variant (TeVes) that couples to the matter field. It's a vector action. Let's focus on the first part of the equation.

$$S_A = -\frac{1}{32\pi G}\int d^4x\sqrt{-\tilde{g}}\left[KF^{\mu\nu}F_{\mu\nu} - 2\lambda(A^\mu A_\mu + 1)\right]$$

MOND is proven to be wrong but let's focus on the vector action in the equation above. Let's multiply it based on FOIL: (a+bi)(a-bi).

$$\left(\frac{1}{32\pi G} + \frac{1}{32\pi G}i\right)\left(\frac{1}{32\pi G} - \frac{1}{32\pi G}i\right) = \frac{1}{512\,\pi^2\,G^2}$$

$$-\frac{1}{512\pi^2 G^2}\int d^4x\sqrt{-(2(g^2))}$$

The value $-1/512\pi^2 G^2$ is interesting because when we take complex conj. pair gravity waves we see the <u>same term</u> $-\pi^2 G^2 512$.

$$2\square^2 2h^{\mu\nu^2} = 2\left(-\frac{\partial^2}{\partial t^2} + \nabla^2\right)2h^{\mu\nu^2} = -\frac{512\pi^2 G^2}{c^8}2T^{\mu\nu^2}$$

We see a <u># correlation</u> "$-\pi^2 G^2 512$" between the complex E-H & Complex gravity waves!

A Linear E-H related to Linear Gravity waves

I don't know the full formulation of this new vector action below but it's a action related to Complex conjugate pair Gravity waves because both have "$-\pi^2G^2512$" in common.

$$-\frac{1}{512\pi^2G^2}\int d^4x\sqrt{-(2(g^2))}$$

Complex conjugate pair Gravity waves based on (a+bi)(a-bi) or i.e. (3+3i)(3-3i)=2(3^2) has the same "$-\pi^2G^2512$" like the vector action.

$$2\Box^2 2h^{\mu\nu^2}=2\left(-\frac{\partial^2}{\partial t^2}+\nabla^2\right)^2 2h^{\mu\nu^2}=-\frac{512\pi^2G^2}{c^8}2T^{\mu\nu2}$$

Based on Wavefunction GR (WGR)

$$G_{2(\mu\nu^2)}=\left(\frac{128\pi^2(G^2)}{c^8}\right)T_{2(\mu\nu^2)}$$

we have complex gravity waves: <u>Real part(mass) Gravity waves</u>+ <u>imaginary mass Gravity waves</u> and <u>complex conjugate pair gravity waves</u>. These 4 waves overlap based on the FOIL method: i.e. (3+3i)(3-3i)=2(3^2). This <u>superimposed Gravity wave</u> is the signature of a Complex conjugate pair big bang.

Gev particle energy for the Linear Graviton

Traditional Gravitational coupling properties is described as the following:

$$GE^2 = (kE)^2/32\pi = (E/M_{planck})^2$$

$$G = K^2/32 = 1/M^2_{Planck}$$

According to linear GR the Graviton is a su-perimposed(superposition) particle. Meaning the Graviton 1st started out as a linear Gravi-ton at the big bang and diffused into separate single eigenstates as the universe grew over time.

$$G_{2(\mu\nu^2)} = (\frac{128\pi^2(G^2)}{c^8})T_{2(\mu\nu^2)}$$

The stress-energy tensor above couples to a superposition(linear) graviton. The superposi-tioned graviton corresponds to different parts of the complex conjugate pair big bang uni-verse: (a+bi)(a-bi) or i.e. (3+3i)(3-3i)=2(3^2). Next we can expand G= K²/32 = 1/M²$_{Planck}$ in complex conj. pair form like gravitons.

The G equation $G = K^2/32 = 1/M^2_{Planck}$
is expressed like $(3+3i)(3-3i)=2(3\wedge 3)$ or like
Linear GR or $\Psi^*\Psi$.

$$\left(\frac{k^2}{32}\pi + \left(\frac{k^2}{32}\pi\right)i\right)\left(\frac{k^2}{32}\pi - \left(\frac{k^2}{32}\pi\right)i\right) = \frac{\pi^2 k^4}{512} = \frac{2}{M^4_{Planck}}$$

Next we can express GE^2 as a complex conjugate pair analogous to $(3+3i)(3-3i)=2(3\wedge 2)$
such as the following.

$$(GE^2+GE^2 i)(GE^2-GE^2 i) = \frac{\pi^2 (kE)^4}{512}$$

<u>Why is this formula special?</u>
"512" in the formula above also shows up in
Complex Gravity waves & a new vector field
Lagrangian. They too have the # 512.

$$2\Box^2 2h^{\mu\nu^2} = 2\left(-\frac{\partial^2}{\partial t^2} + \nabla^2\right)^2 2h^{\mu\nu^2} = -\frac{512\pi^2 G^2}{c^8} 2T^{\mu\nu^2}$$

$$-\frac{1}{512\pi^2 G^2}\int d^4x\sqrt{-(2(g^2))}$$

The number "512" numerical correlations between formulas suggests a Linear nature to
General Relativity and Gravitons.

512 and the
new linear Ricci curvature tensor

The Ricci Curvature tensor in four dimensions has 256 components. If we look at the Ricci Curvature tensor in linear GR the number of component is 512.

$$R_{2(\mu\nu^2)} - \frac{1}{2}g_{2(\mu\nu^2)}(2R^2) = (\frac{128\pi^2(G^2)}{c^8})T_{2(\mu\nu^2)}$$

The linear Ricci tensor $R_{2(uv^2)}$ means $2 \times ((R_{uv}$ curvature tensor(256 components))^2). If the linear GR formula has 512 components to it's Ricci curvature tensor then that may shed insight as to why we see the number "512" in the Complex gravitational waves and a new vector field Lagrangian.

$$2\Box^2 2h^{\mu\nu^2} = 2\left(-\frac{\partial^2}{\partial t^2} + \nabla^2\right)^2 2h^{\mu\nu^2} = -\frac{512\pi^2 G^2}{c^8} 2T^{\mu\nu^2}$$

A vector field Lagrangian from a TeVes(Tensor–vector–scalar gravity) expressed as complex conjugate pair has 512 in it.

$$S_A = -\frac{1}{32\pi G}\int d^4x\sqrt{-\tilde{g}}\left[KF^{\mu\nu}F_{\mu\nu} - 2\lambda(A^\mu A_\mu + 1)\right]$$

$$-\frac{1}{512\pi^2 G^2}\int d^4x\sqrt{-(2(g^2))}$$

234

The value of 512 appears in the following GR and GR related formulas.

Linear GR has <u>512 components for</u> $R_2(_{uv}{}^2)$. R_{uv} only need 20 components but it can have up to 256 components: 2*256=512 components.

$$R_{2(\mu\nu^2)} - \frac{1}{2} g_{2(\mu\nu^2)}(2R^2) = \left(\frac{128\pi^2(G^2)}{c^8}\right) T_{2(\mu\nu^2)}$$

Complex Gravitational waves has the # 512.

$$2\Box^2 2h^{\mu\nu^2} = 2\left(-\frac{\partial^2}{\partial t^2} + \nabla^2\right)^2 2h^{\mu\nu^2} = -\frac{512\pi^2 G^2}{c^8} 2T^{\mu\nu^2}$$

A new type vector field Lagrangian has the # 512 in it.

$$-\frac{1}{512\pi^2 G^2}\int d^4x \sqrt{-(2(g^2))}$$

The equation related to graviton energy coupling and temperature has the #512 in it.

$$\frac{\pi^2(kE)^4}{512}$$

Linear GR based on complex conjugate pair (a+bi)(a-bi) stress energy multiplication analogous to $\Psi^*\Psi$ quantizes GR and makes it compatible to the linear physics of QM $\Psi^*\Psi$.

Test 3: Numerical Relativity tests

Linear GR based on a linear Graviton

$$R_{2(\mu\nu^2)} - \frac{1}{2}g_{2(\mu\nu^2)}(2R^2) = \left(\frac{128\pi^2(G^2)}{c^8}\right)T_{2(\mu\nu^2)}$$

describes a universe having a complex conjugate pair big bang were E=2c⁴m² =(mc²+(mc²) i)(mc²-(mc²)i) opposed to a matter-antimatter big bang. The LINEAR Gravity waves from this complex conjugate pair big bang can be simulated to see if they match anything in the WMAP.

Linear Gravity waves

$$2\Box^2 2h^{\mu\nu^2} = 2\left(-\frac{\partial^2}{\partial t^2} + \nabla^2\right)^2 2h^{\mu\nu^2} = -\frac{512\pi^2 G^2}{c^8}2T^{\mu\nu^2}$$

The Graviton, Gravity waves, etc. based on Linear GR can be simulated in a computer. The idea is that we need nonlinear GR to describe Classical physics like space curves and Gravity waves. However, for QG effects we need linear GR. The true form of the Graviton may be a linear Graviton and Linear Gravity waves may explain mysteries in the WMAP.

236

Numerical Relativity test for expansion

Let's separate the linear GR formula

$$R_{2(\mu\nu^2)} - \frac{1}{2}g_{2(\mu\nu^2)}(2R^2) = (\frac{128\pi^2(G^2)}{c^8})T_{2(\mu\nu^2)}$$

as (a+bi) and (a-bi). We had equal parts real & imag. energy analogous to (3+3i)(3-3i)=2(3^2).

TACHYONS REPLACE LAMBDA Λ

Our complex universe started with equal parts matter and imaginary matter.

$$R_{\mu\nu+\mu\nu i} - (\frac{1}{2} + \frac{1}{2}i)g_{\mu\nu+\mu\nu i}R + Ri = \frac{8\pi G}{c^4} + (\frac{8\pi G}{c^4}i)T_{\mu\nu+\mu\nu i}$$

The imaginary mass(energy) is causing the universe to expand and accelerate because Tachyon FTL physics works in reverse compared to slower than light mass physics. Below is our complex conjugate pair universe.

$$R_{\mu\nu-\mu\nu i} - (\frac{1}{2} - \frac{1}{2}i)g_{\mu\nu-\mu\nu i}R - Ri = \frac{8\pi G}{c^4} - (\frac{8\pi G}{c^4}i)T_{\mu\nu-\mu\nu i}$$

Complex universe curves
(x²,x²/x²,1/x² & -x²,-x²/x²,-1/x²)

We can't see
Tachyons but we
see it's anti-gravity
x^2,-x^2
space curve effects!

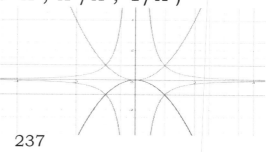

237

<u>Numerical Relativity for WGR galaxies</u>
(WGR is also known as Linear GR)
Wavefunction GR(WGR) puts FTL GR on
equal footing with FTL Special Relativity. For
instance, when objects travel FTL in Special
Relativity the math shows how the mass becomes imaginary and time reverse. Essentially FTL physics reverses. Tachyons can't slow
down to C just like real mass can't reach C.

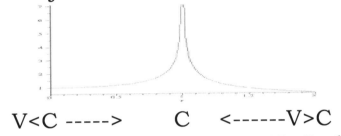

V<C ----->   C   <------V>C

<u>GR at V>C correlate to SR V>C physics</u>
The idea is that for FTL spaceflow below the
Event horizon of a Blk hole the physics correlate to FTL SR physics. Meaning space flow
reverses, time reverse were you have a imaginary mass White hole underneath a blk hole!

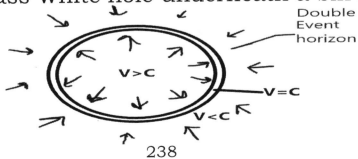

238

Not MOND but added Tachyon anti-gravity

When the $1/\sqrt{r}$ decreasing rotational curve from real mass overlap the $\sqrt{r}$ increasing rotational curve from the white hole inside then we have a $\sqrt{r}/\sqrt{r}=1$ flat rotational curve for the galaxy.

Opposing rotational curves causes the flat rotational curves. The real & imag. mass(gravity) in our universe causes our universe to be flat. The inflationary curve x^2 from imag. mass anti-gravity overlaps the deflationary curve $1/x^2$ from real mass. Their overlap $x^2/x^2=1$ causes the universe to be flat. Side Graph is of complex conj. pair universe curves. Tachyons play a role in flat rotation curves and universe expansion.

There is no dark matter material

The flat rotational curves & flat geometry of the universe is due to a overlap between gravity(mass) & anti-gravity(imag mass). Computer simulation can test it & the Bullet cluster.

The Bullet cluster simulation

Based on Linear GR the flat rotational curve for galaxies is due to an overlap of the rotational curves between the internal imaginary mass(energy) white hole and the outer black hole holes. Their opposing rotational curves creates flat rotational curves. The premises for the argument is that FTL spaceflow below the event horizon behaves like FTL Special Relativity. Meaning FTL spaceflow geodesics have imaginary energy(matter) physics. The laws of physics reverse just like FTL Special Relativity. Space inside a blackhole isn't cascading down to a central singularity but space is flowing away from a central point. This white hole inside a blackhole has a external rotational curve of x^2 which is the reverse of the $1/x^2$ rotational curve from the black hole. The two curves overlap to form a flat rotational curve. It's not dark matter.

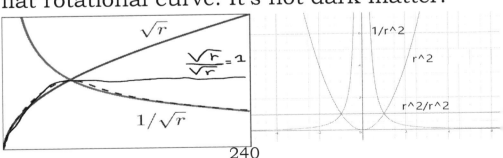

240

The new rotational curve model is based on a single galaxy with a central black hole inside.

A complicated Bullet cluster simulation

The bullet cluster is based on a collection of two galaxies merging. We would need to take the following into account for a simulation:

1. Count(approx.) all the individual black holes within each of the cluster of galaxies.

2. Model and track how all the individual galaxies interact with each other based on the new flat rotational curve model.

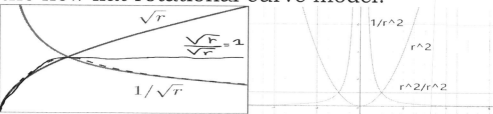

3. The new model suggests it's not "Dark matter" but a complicated interaction of the individual galaxies flat rotational curves into one overall interaction to form the bullet cluster pattern.

Predictions for Linear GR (LGR)
also known as Wavefunction GR(WGR)

1. Linear(Superimposed) Graviton: mass of the Graviton is $\pi^2/128$, similar to the W Boson (1/128). This expanded version of GR describes Gravitons as mediating in wave-like superimposed particle form rather than single eigenstate point-like Graviton form.

$$G_{2(\mu\nu^2)} = \left(\frac{128\pi^2(G^2)}{c^8}\right)T_{2(\mu\nu^2)}$$

"This Linear Graviton could be detected in a particle accelerator. Gravitons started out linear at the big bang & diffused into single eigenstate GR Gravitons overtime "
The superimposed Graviton from WGR is mediating in a superposition state. We don't have point-like changes but wave-like(superimposed exchanges). Compared to the W-Boson(1/128) the coupling constant of the superimposed Graviton is $\pi^2/128$.

$$\alpha = \frac{\pi^2}{128} \qquad \begin{array}{l} \alpha \approx 1/128 \\ \text{at } Q^2 \approx m^2(W) \end{array}$$

2. New Primordial linear Gravity waves: WGR describes a complex conjugate pair big bang. We can detect it's Primordial Gravity waves.

$$G_{2(\mu\nu^2)} = (\frac{128\pi^2(G^2)}{c^8})T_{2(\mu\nu^2)}$$

$$2\Box^2 2h^{\mu\nu^2} = 2\left(-\frac{\partial^2}{\partial t^2} + \nabla^2\right)^2 2h^{\mu\nu^2} = -\frac{512\pi^2 G^2}{c^8}2T^{\mu\nu^2}$$

Quantum fluctuations or LINEAR Gravity waves in the WMAP?

The WMAP is conjectured to be based on cosmic inflation. What if they are superimposed Gravity waves from a complex conjugate pair big bang. We can test to see if it is true.

3. Numerical Relativity tests: Before experiments, WGR simulations can be conducted.

Negative mass in linear GR(LGR)

According to linear GR we had a complex conjugate pair big bang analogous to $\Psi^*\Psi$. We had equal parts real mass & imag. mass.
$$(a+bi)(a-bi) \text{ ,i.e. } (3+3i)(3-3i)=2(3^2)$$
On a complex plane it corresponds to a complex point$(a+bi)$ and it's complex conjugate pair point$(a-bi)$.
Interestingly if have negative mass or negative pressure then it would exist as a different set of complex conjugate point pairs.
$$(-a+bi)(-a-bi)$$
i.e. $$(-3+3i)(-3-3i)= 2(3^2)$$

Negative complex big bang

Just as we had a complex conj. big bang their could be a negative complex conj. big bang.

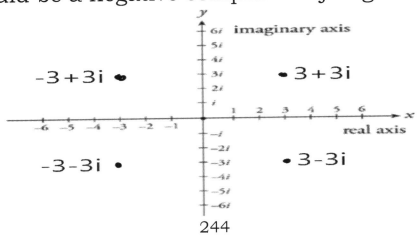

244

The Linear GR landscape(LGR)

If we introduce negative mass in LINEAR GR then it will exist as a separate set of complex conjugate pair points.

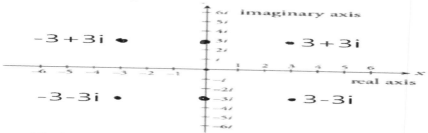

What if the shape of negative mass is negatively curved? In our universe the particles are spherical, large scale planets, moons and stars are spherical. Possibly our universe may collapse into a big crunch which would make the shape of our universe spherical. What if negative mass is completely negative, including it's universe geometry? Also the imaginary axis(i.e. 0+3i, 0-3i) is analogous to flat Euclidean geometry with a cylindrical curved universe on it. It's an eternal universe with no beginning and no end. The geometric landscape of Linear GR on the Quantum & cosmological scale is Non-Euclidean.

Complex symmetries

Linear GR has multiple symmetries in the complex plane.

1. <u>Complex conj. pair symmetry (a+bi)(a-bi)</u>. What's interesting is that real particles are their own complex conjugates. The real particles in our universe don't have partner particles. Their is no SUSY. Only the tachyons(Dark Energy) have partners(+bi,-bi).

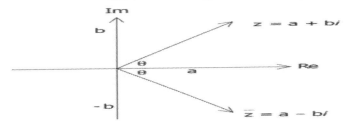

Just as we need matter-antimatter or virtual particles-anti virtual particles for the QFT of the EM, weak, strong,etc. fields we need complex conjugate pairs for the QFT for gravity. The complex values below are <u>single point particles represented in the complex Quantum field plane</u>.

(1/16π+1/16πi)

(1/16π - 1/16πi)

2. <u>Negative complex symmetry</u>: We can have a negative complex big bang (-a+bi)(-a-bi) that's opposite of an (a+bi)(a-bi) complex big bang. Real mass has spherically shaped particles. The imag. axis is analogous to a Cylindrical universe(no beginning/end) with scalar field point particles. The negative side has a 3D hyperbola curved universe(finite) with negative mass particles shaped like mini 3D hyperbolas. It's big bang started as a ring(base on 3D hyperbola) and will end in a ring.

3. <u>Vacuum symmetry</u>. The X-axis is the vacuum state. We have positive mass spin & negative mass spin particles from the X-axis. The origin(Spin 0) is a scalar field. We can have 0+3i,0-3i scalar point particles in a cylindrical universe and 3+3i,3-3i/-3+3i,-3-3i complex point particles in a negative/positive curved universe. We have a reflection between the (+) vacuum and (-) vacuum on the X-axis. The origin(0) is a scalar field between the (+)positive and (-)negative vacuums.

Negative mass universe

(Negative complex big bang$(-a+bi)(-a-bi)$)

If we have a negatively curved universe were the subatomic particles are 3D hyperbolas then we have atomic nucleus of 3D hyperbola particles(i.e. 3D hyperbola hydrogen negative mass proton and a 3D hyperbola shaped negative mass electron orbiting). If that is true then the universe will have large scale moons, stars shaped like 3D hyperbolas. The overall shape of the negatively curved universe is a 3D hyperbola. This logic is based on the fact that our universe has spherical particles, spherically shaped moons, suns and a possible spherically shaped geometry(Big crunch). If we extend the complex GR cosmological model to the rest of the complex plane when we can have a negative mass complex conjugate pair big bang $(-a+bi)(-a-bi)$. Everything may be purely negative including the shape of particles. The hyperbola universe could be made of finite size 3D hyperbola size particles. It's particle interaction and quantum behavior will cause a large scale interesting negatively curved macro landscape.

The negative mass Atom

The idea is that a negative mass universe is totally negative. Meaning not only is it's mass negative, but it's geometry is negative on the quantum and cosmological scale. The shape of it's atoms are 3D hyperbolas. Below is a simplified picture of a negative mass atom. The 3D hyperbola electron would exist in a "cloud" but for simplicity sake I am presenting it as a circular orbit.

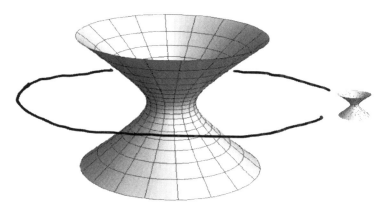

Imagine an atomic nucleus were the 3D hyperbola protons and neutrons are arranged in a certain configuration. On the macro scale the shape of suns and planets are shaped like 3D hyperbolas. The overall geometry of the negative mass universe is a 3D hyperbola.

The negative standard model

The negative mass universe will have it's own standard model of physics(reflection of our positive mass standard model). I am assuming electromagnetic radiation will still travel at the speed of light. However, the configuration of the EM waves will be different from that of a moving spherical charged particle.

A ringed big bang and end

From our spherical universe perspective it started from a superposition point(big bang) and will end in a superposition point(big crunch). Just as our spherical universe has a finite time(big bang to big crunch), I'm assuming a 3D hyperbola universe has a finite time span. Meaning it started out as a "Big ring". That big ring is the base of a 3D hyperbola. The circular ring universe will contract at it's equator and then expand again towards a finite big ring end. The 3D hyperbola universe has a finite start and finite end like a spherical universe. It's only the cylindrical scalar field universe that has no start or end.

Non-Euclidean standard model of particle physics

Positive mass particles having spherical shapes(& complex conj. pairs) and negative mass particles having 3D hyperbola negative curved shapes(& complex conj. pairs), & Scalar particles(& complex conj. pairs: i.e. 0+3i,0-3i) complete the particle landscape. The non-Euclidean standard model of particles comes from a complex spacetime QFT. The real axis is the vacuum ground state. As the field oscillate up and down in the complex plane it forms particles once the vacuum (ground state) reaches a certain energy level. This complex field can create our real mass universe(spherical particles & complex conjugate spherical particles), scalar field particles associated with a cylindrical universe with no beginning and not end and the Q.F. can form negative 3D hyperbola particles associated with a hyperbola universe. Non-euclidean particles physics dismisses String theory.

The Linear GR(LGR)
particle physics landscape

The Feynman diagram for Linear GR is based on a complex spacetime. Conjecturally all particles(electrons, EM field, etc.) can form from complex spacetime Quantum fields. Also, if we look at the complex landscape of LGR & relate the real axis to particle spins it directly support a Multiverse landscape. Once a particle form from a Q. Field it has spin.

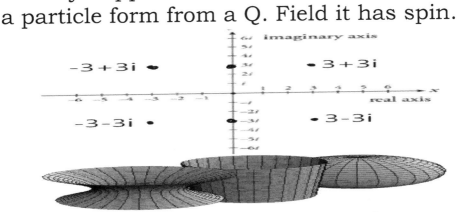

Particle Spin number line

$$\cdots\ -2\ \ -\frac{3}{2}\ \ -1\ \ -\frac{1}{2}\ \ 0\ \ \frac{1}{2}\ \ 1\ \ \frac{3}{2}\ \ 2\ \ \frac{5}{2}\ \ 3\ \ \frac{7}{2}\ \ 4\ \cdots$$

-P.U.   **Negative reflection universe**   **(Higgs universe)**   **Our universe**   **Parallel universe**   P.U.

The Multiverse "spin" equation

All the particles in our universe is captured in the 1st 4 spins. Higgs is Spin 0. Conjecturally parallel universes are also based on 4 fundamental particle spins. It's like our universe but shifted along the "spin" number line.
Our universe total spin states(universe A):

$$(1/2)+1+(3/2)+2=\underline{5}$$

Parallel universe total spins (universe B)

$$(5/2)+3+(7/2)+4=\underline{13}$$

Another parallel universe(universe C)

$$(9/2)+5+11/2+6=\underline{21}$$

Another parallel universe(universe D)

$$(13/2)+7+15/2+8=\underline{29}$$

Another parallel universe(universe E)

$$(17/2)+9+(19/2)+10=\underline{37}$$

The Parallel universe group spin(P) sequence 5,13,21,29,37... is based on the equation!

$$P_n = 8n-3$$

Negative mirror universes are $-P_n = 3-8n$
The "Multiverse equation" $P_n = 8n-3$ above describes our multiverse! The letter "n" represent a different universe: n=1,$\underline{5}$(spin sum);n=2,$\underline{13}$;n=3,$\underline{21}$;n=4,$\underline{29}$,etc.

The Multiverse Equation

The Multiverse equation based on particle spin groups could be a new area of physics called <u>Multiverse Particle physics</u>.

$$P_n = 8n-3$$

Our universe total spin states(universe A):

$$(1/2)+1+(3/2)+2=\underline{5}$$

Parallel universe total spins (universe B)

$$(5/2)+3+(7/2)+4=\underline{13}$$

Another parallel universe(universe C)

$$(9/2)+5+11/2+6=\underline{21}$$

Another parallel universe(universe D)

$$(13/2)+7+15/2+8=\underline{29}$$

Another parallel universe(universe E)

$$(17/2)+9+(19/2)+10=\underline{37}$$

<u>Example: Physics of Universe B</u>

Just like our universe can only have the spin range & sum of $1/2+1+(3/2)+2=5$, <u>Universe B</u> can only have the following spin ranges.

1. Electron spin is $5/2$
2. Photon spin is 3
3. Unknown Fermion particle spin $7/2$
4. Graviton Spin 4

Multiverse Supercomputer simulations

Based on the Multiverse equation below.

$$P_n = 8n-3$$

it may be possible to model parallel universes based on the 4 fundamental particle spin
states for that particular universe. Supercomputer simulations or virtual reality or gaming
software, MV movies can create a true multiverse experience based on the 4 fundamental
spin states for particular universes. It will be
a true physics engine software model for parallel universes. The software can be tested by
using the 4 fundamental spins for our universe:$(1/2)+1+(3/2)+2=5$ and see if it matches
our observable universe. Afterwards the software can shift to other particle spin groups to
show how the laws of physics will behave for
that universe.

The Multiverse(MV) equation

Just as we have the Drake equation

$$N = R^* \cdot f_p \cdot n_e \cdot f_l \cdot f_i \cdot f_c \cdot L$$

we now have the MV equation: $P_n = 8n-3$.

Additional math & science discoveries

"The formula of love"

I've discovered(God revealed to me) a mathematical interpretation of the Holy Bible verse(Genesis 2:24).

Genesis 2:21-24 King James Version (KJV)

21 And the Lord God caused a deep sleep to fall upon Adam, and he slept: and he took one of his ribs, and closed up the flesh instead thereof;

22 And the rib, which the Lord God had taken from man, made he a woman, and brought her unto the man.

23 And Adam said, This is now bone of my bones, and flesh of my flesh: she shall be called Woman, because she was taken out of Man.

24 Therefore shall a man leave his father and his mother, and shall cleave unto his wife: and they shall be one flesh.

256

Mathematical interpretation of "They shall be one flesh" Genesis 2:24

Literally, the Bible verse of "They shall be one flesh" doesn't make any sense. However, from a purely mathematical perspective it's describing a equation. For example,

$$3/1 \times 1/3 = 1$$
$$5/1 \times 1/5 = 1$$
$$8/1 \times 1/8 = 1$$

From a purely mathematical perspective the bible verse of "They shall be one flesh" is stating that men and women are inversely symmetrical like 3/1 and 1/3. Multiplied together they equal 1. Just as a man and woman's body is physically reverse (i.e. his private parts protrude out of his body while her private parts protrude inward) their mental and emotional configuration are reverse as well. Meaning women think and feel reverse of men like their physical bodies are inverted.

Men and women shouldn't treat each other as the opposite sex but the inverted sex. Men and women don't think or feel the same way. They should treat each other inversely.

What is the difference between the opposite sex and inverted sex?

Men and women were thought to be the opposite sex were their equal and opposite like matter and antimatter.

However, if you look at the physical shape of a man and woman's body they are inverted(reverse0 rather than opposite analogous to a sphere and hyperbola. A man has budging oval spherical type shape while a woman has a general hour glass shape. The relationship difference is analogous to a sphere and 3D hyperbola.

The difference between a man and woman is analogous to 3/1 & 1/3 rather than +3 and -3. The inverted relationship isn't stating women are lesser or inferior to men but is inversely symmetric to how men are.

The inverted gender sex

Just as a woman and man's body were inversely symmetrical so are their emotional make up. For example, a man genitals are on the outside his body while a woman's genitals are inside her. The idea is that a man and woman's emotional make up is inversely symmetrically like their physical comparison is inversely symmetrical.

Beauty in symmetrical love

The formula of love describes inverse symmetry between men and women. That inverse symmetry is the beauty in love. She's very good looking & the guy has status, height,etc. Problems arises when asymmetric people want to be symmetric with beautiful people because they think too highly of themselves. Men should realize beautiful women can be very nice if she likes a guy or a high status guy will cherish/marry a woman if he likes her. When people are always frustrated and being rejected by certain attractive people they need to focus on people who want them. It's not about lowering standards but being symmetrical to someone who wants you.

The Love function from the Holy Bible

Genesis 2:24 (they shall be one flesh) can be interpreted as a mathematical equation similar to 4/1 x 1/4 = 1 or 5/1 x 1/5 =1.

$$\frac{\male}{1} \times \frac{1}{\female} = 1 \text{ flesh}$$

"1 flesh" is mathematically interpreted as men and women relating inversely to harmonized together as a good couple. The reason why men are the larger value is due to physical size difference between men and women. The point is that when men and women interact in a relationship they should not treat each other the same. He don't feel or think the same as she does and vice versa. Amazingly they'll have to love each other by faith and not by sight. Meaning even though he doesn't feel or relate to the type of love she wants he should rely on faith in the bible version Genesis 2:24 she likes that type of treatment. The same applies to her treating him like he wants where she don't relate to it.

General inverted love pattern

Men	women
Talk straight to point	Don't know the point until finish talking
Straight to shopping	Don't go straight to finish shopping but enjoys the experience
Rushing towards sex	Takes time to build up to sex to get into the mood
Action movies, sports activities together	Romance (dates, vacations)
House can be messy	Require a clean house
Disciplining parenting	Emotionally nur- turing parenting
Love by providing	Love by spending time together

Men	women
Attracted to her beauty	Attracted to his Money, security, status excitement,Hypergamy, bad boys (substitute for wealthy or financially secure men)
He don't care to share with her what he's thinking. (He feels if he wanted her to know then he would tell her)	She want to know what he's thinking
Need to feel respected	Need to be cherished (embraced, attended to)
Wants honesty (not deceiving him: using him for $$)	Wants honesty (not pretending he loves her: using her)
He wants her to try to keep in shape	She wants him to be groomed, clean cut
Don't want to be smothered	She want him to be confident

Reinforcing vs undermining

If men and women relate inversely then they reinforce each other like light constructive interference. If men & women treat each other same(she give him what she want & vice versa) they undermine each other & grow apart.

Love by faith & not by sight

Men don't feel the type of love she want and vice versa. To harmonize as a couple he has to love by faith were he gives her the type of love she need even though he don't need it or feel the same way & vice versa. For men and women to harmonize they need to love by faith and not by sight. She needs to understand what stimulus him and make him emotionally happy isn't the same as what she want and vice versa. Men & women need to love by acting on faith that their partner want inverted love. The Formula of love also highlights only certain men and women are compatible: 3/1 x 1/3=1 were 5/1x1/6 don't equal 1 based on Genesis 2:24. Love is unique!

$$\frac{\male}{1} \times \frac{1}{\female} = 1 \text{ flesh}$$

A Voynich Manuscript discovery

According to Wikipedia the Voynich Manuscript is "an illustrated codex hand-written in
an unknown writing system".
No one can decipher its hand-written text but
what makes the Voynich manuscript so interesting are the images. Some of the images
have been identified but other images are a
puzzle. I believe I have found a new discovery
in one of the images. <u>Look at the pattern of</u>
<u>the plant on the leftside</u>.

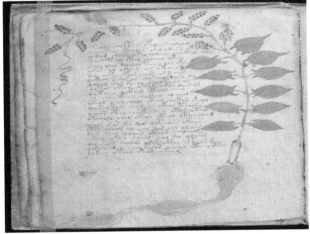

Notice how it resembles the side profile of a
human face: Forehead, nose, lips, beard &
maybe a mustache. Also a tongue may be
sticking out between the lips. It's not out of

the question that a Voynich Manuscript plant has a side profile of a human face. In another area of the Voynich manuscript it has a leaf shaped like a Elephant head. <u>Look at the lower right side of the page below</u>.

<u>Forensic Artist experiment</u>

If the Voynich Manuscript has a flower that has the side profile of a human face then it's possible a Forensic Artist can extrapolate a 3D face called the "Voynich Face". The Face is interesting because it appears to have it's tongue sticking out. That may have had a meaning during the middle ages. The Face can undergo Facial recognition to see if it matches a human face or statue face.

20. <u>Other book by me</u>

If you've been enlighten by this book then try out one of my other unique science related books.

<u>Prayer and Parallel Universes! Could there also be parallel Heavens?</u>

(Book cover image on next page)

Have you ever been curious about the science of the multiverse? It is on the fringe of theoretical physics and a favorite theme in the science fiction genre. I've went a step further and explored the possibility that if parallel universes are real then from a religious perspective parallel Heavens should exist as well. To bring parallel universes into something relatable I speculated that our dreams that we have while we are asleep at night are glimpses into Parallel Earths, we should pray over. <u>It's not adding or subtracting from our bible.</u> Parallel Earth prayer states other universes via dream interpretation have their own Holy Bible version, God, Heaven like people and Earths have parallel versions. All parallel versions are one like supercomputer oneness.

<u>My other book available on Amazon</u>
Below is my other new and exciting book. It explores a new possibility on the science of parallel universes. Available on Amazon.

<u>Closing remarks</u>

As God blesses me with new mathematical insight I share it with the world.

1 Corinthians 1:27 King James Version (KJV)

27 But God hath chosen the foolish things of the world to confound the wise; and God hath chosen the weak things of the world to confound the things which are mighty;

Printed in Great Britain
by Amazon

50215439R00161